a book from Prince Edward Arts Council,
your County Library and Books & Company

fires!

• • •

true stories from the edge

Tanya Lloyd Kyi

fires!

Ten stories that
chronicle some
of the most
destructive fires
in human history

ANNICK PRESS

TORONTO + NEW YORK + VANCOUVER

Annick Press Ltd.

All rights reserved. No part of this work covered by the copyrights hereon may be reproduced or used in any form or by any means—graphic, electronic, or mechanical—without the prior written permission of the publisher.

We acknowledge the support of the Canada Council for the Arts, the Ontario Arts Council, and the Government of Canada through the Book Publishing Industry Development Program (BPIDP) for our publishing activities.

Edited by Pam Robertson
Copyedited by Naomi Pauls
Cover art by Scott Cameron
Design by Irvin Cheung/iCheung Design

Cataloging in Publication Data

Lloyd Kyi, Tanya, 1973–
    Fires! / by Tanya Lloyd Kyi.

(True stories from the edge)
Includes bibliographical references and index.
ISBN 1-55037-877-5 (bound). — ISBN 1-55037-876-7 (pbk.)
    1. Fires—History—Juvenile literature. 2. Disasters—History—Juvenile literature.
I. Title. II. Series

TH9448.K95 2004   j363.37'09   C2004-902202-4

The text was typeset in Bembo.

Published in the U.S.A. by	**Distributed in Canada by**	**Distributed in the U.S.A. by**
Annick Press (U.S.) Ltd.	Firefly Books Ltd.	Firefly Books (U.S.) Inc.
	66 Leek Crescent	P.O. Box 1338
	Richmond Hill, ON	Ellicott Station
	L4B 1H1	Buffalo, NY 14205

Printed and bound in Canada

Visit oue website at **www.annickpress.com**

Contents

Introduction
The Myths and Reality of Fire

IT WAS 1812, and the French emperor Napoleon Bonaparte was battling his way through Russia. The French and Russian armies met in bloody clashes, leaving bodies strewn across the countryside. But Napoleon had gathered the largest force in history—there seemed to be no way to stop him. Soon the Russian army began to collapse. As the French marched closer and closer to the great city of Moscow, the Russians knew their capital would soon be conquered.

Russian aristocrat Count Feodor Rostopchin packed his most precious possessions and prepared to flee. He refused to stay and be ruled by Napoleon. Just before dawn, he roused the Russian army officers who were guarding his country estate. He pressed flaming torches into their hands and gave them their instructions. The men were shocked, but they knew better than to argue with Rostopchin. Within minutes, they were setting fire to the count's thick velvet curtains, ornately carved wooden chairs, and priceless works of art. When the mansion was ablaze, the count posted a note on the front gate: "I would rather set my home on fire than have it polluted by the French. I abandoned to you my two houses in Moscow . . . Here you will find only ashes."

Though most Russian aristocrats fled along with Rostopchin, Napoleon never succeeded in ruling Moscow. Shortly after he arrived with his army, fire consumed most of the city. It began in a Chinatown market and quickly spread. Soon businesses were blazing, rows of colorful houses became streams of fire, and towering flames surrounded Moscow's most important government buildings. The inferno roared through the dark of a Russian night, transforming all of Moscow into a raging bonfire. By the time rain doused the blaze four days later, the city was reduced to rubble.

Many people believed the Russians followed Rostopchin's lead and started the blaze so the French couldn't live there. Others believed that drunk French soldiers accidentally spread the flames. Whatever the cause, Napoleon was left without food and shelter for his men. With victory wrenched from his grasp, he was forced to march back to France through the harsh cold of a Russian winter.

The First Sparks

Like many stories of fire, Rostopchin's tale is full of drama and destruction, danger and desperation. Fire is a powerful force and brings out powerful emotions. Today when we see a building ablaze or a forest in flames, we are filled with both awe and fear. In the past, people were so fascinated by fire that they wove it into many of their myths and legends. They tried to explain where flames came from and how people first began to use fire.

The ancient Greeks believed Zeus, the most powerful of the gods, controlled the secrets of fire. Zeus was angry with Prometheus, the creator of humans, so Zeus refused to teach people how to use fire. But Prometheus outsmarted Zeus. He scaled Mount Olympus and touched a torch to the sun. Then he carried the burning wood back down to the people and taught them how to use the flames.

According to the Cherokee people of North America, the very first fire was started by lightning, in the hollow of a sycamore tree. Humans saw the flames and wanted to use them, so several creatures tried to get close to the blaze and bring out an ember. Raven tried, but his feathers were scorched by the heat. Screech Owl tried and burned his eyes. Snake crawled in, only to become black and twisted. Finally, Water Spider spun a bowl from her web and carried out a single coal. From this coal, people started their first fire.

In South America, people say that Jaguar originally controlled fire. He hunted with bows and arrows and used fire to cook his food. One day Jaguar met a man in the jungle and invited him home. The man saw the weapons and the fire and he was jealous of Jaguar's powers. He killed Jaguar's wife so he could steal the bows and arrows and the flame. After that day, people used fire, but they lived in constant fear of Jaguar's revenge.

Centuries after storytellers spun these myths, people continue to rely on the power of fire—to heat their homes, to cook their food, and sometimes to clear their land. And they face the same problem that early users of fire discovered. They can create fire, they can use it—but they can't necessarily control it.

Fire Power

British soldiers, Indian housewives, street urchins, and Portuguese traders mingled among Bombay's market stalls on February 17, 1803. Vendors offered vibrantly colored tropical fruit, teak carvings, squawking live chickens and ducks, parchment and quills, freshly fried breads, and hand-painted fabrics. The air was filled with a hundred different smells and a jumble of sounds. No one noticed a single overturned candle.

Suddenly the crackle of wood and the whoosh of burning canvas punctuated the noise. A market stall was on fire! Vendors rushed to douse the flames but a strong wind swept the blaze from one stall to the next, then to the next. Soon the entire bazaar was a deafening furnace. The crowds that had gathered to gawk at the wares now pushed and shoved to escape the narrow lanes.

British military staff responded quickly, pulling two fire pumps toward the scene. They were beaten back by the heat. A wall burned and collapsed, killing two captains and their crews.

Incinerating the wooden stalls and shacks of local families, the fire tore along the waterfront toward the army barracks. For the entire day, hundreds of people rushed to help, carrying buckets of water, operating hand pumps, or tearing down buildings in the fire's path. Still the flames advanced. Temples burned. Families rushed in panic toward the docks or farther into the city, frantically searching for safety.

Finally soldiers devised a desperate plan. They turned their cannons on the city itself, bombarding a row of homes and businesses until a great swath lay empty of fuel for the fire. Soldiers and local residents held their breath as the flames approached this last line of defense. For a moment, the fire swelled as if it would overtake the firebreak, then it shrank and finally died. There were tears of relief. The soldiers had succeeded. Still, a third of the city lay in smoking ruins, and more than 450 houses had been destroyed.

Battling the Blaze

Today fires rarely threaten entire cities. When a building fills with flames and smoke, alarm bells sound and sprinkler systems activate. Modern firefighters respond in minutes, connect to municipal water supplies, and attempt to douse the flames before they can spread. The tools these rescuers use—from the alarms that first alert them to the pumps that help shower the flames with water—didn't exist when the stalls of the Bombay bazaar burned in 1803.

Fire was a constant threat in cities in centuries past. Neighborhoods were packed with wooden buildings on narrow streets, and the only tool people could use to protect their homes was the water bucket. There were a few obvious problems. It was difficult to carry enough water toward a blaze. In the case of a big fire, it was almost impossible to carry buckets through the smoke and heat to the flames themselves. A few small hand pumps mounted on wheels allowed residents to use larger amounts of water, but even the pumps seemed useless against a building or an entire city block on fire.

In the 1670s, a Dutch engineer invented the fire hose. Made of leather, the sturdy tube allowed early city dwellers to transport water over longer distances and send it shooting closer to the flames. Longer ladders and better water pumps later helped people use the hoses more efficiently.

Unfortunately, not everyone knew how to use this equipment. Often people panicked when faced by a fire and didn't use the hoses or the pumps properly. In some cities, officials decided to recruit strong men and train them to fight fires—and that's how fire departments began. In the United States, Benjamin Franklin organized men to serve as full-time firefighters in Philadelphia in 1740. City residents paid to be put on the list of homes the organization would protect. Edinburgh, Scotland, created the first official city fire department in 1824, and London established one soon after.

These newly established fire departments practiced responding quickly to alarm bells. A new invention helped—the telegraph. In the mid-1800s, building owners began to wire their buildings to alarm boxes in the nearest fire stations. Now they could directly signal the firefighters. They no longer had to wait for people to see the smoke or hear the bells and carry the news.

Of course, there have been countless important fire-fighting inventions since the telegraph. Today we have smoke alarms, house alarms, and sprinkler systems. Modern fire crews are equipped with two-way radios, cellular phones, truck-mounted ladders, heat-resistant clothing, and first aid devices. With our advanced technologies and our well-trained firefighters, we have little worry that our cities or towns will burn. We even feel safe in our homes when we see the steady light of our smoke detector shining down from the ceiling. But what if a fire begins where there are no alarm bells or sprinkler systems? Can firefighters still respond quickly and effectively?

Forests in Flames

Lightning strikes, or a burning cigarette butt is tossed from a car window. Suddenly the dry underbrush is in flames, then the hillside, then the entire forest. Some fires help forests renew themselves, fertilizing the soil and allowing plants more room to grow. But other forest fires attack valuable timberland or approach the edges of cities. When this happens, specially trained crews pour in to help control the blaze.

In 1994, hot dry winds swept more than 800 bushfires through Australia's Hunter, Blue Mountain, and Sydney regions. Airplanes buzzed overhead, providing constant surveillance. Helicopters dipped huge buckets into lakes, ponds, and even local swimming pools, trying to cool the edges of the fires. Hundreds of bulldozers roared forward and back, creating wide tracks of dirt meant to slow the fire's spread.

Thousands of volunteer firefighters worked to evacuate the residents of nearby towns and limit property damage. There was no stopping the bushfires themselves. As the flames heated massive eucalyptus trees, the leaves gave off flammable vapors. Huge banks of smoke were occasionally broken by exploding fireballs.

Again and again, Australian firefighters doused a row of homes in a Sydney suburb with water and fire retardant. Just a kilometer (about half a mile) away, the flank of a massive bushfire sent out searing waves of heat. The residents of the suburb had already fled. From emergency shelters, they watched the hourly updates on the news. The fire grew closer and closer to their homes. Then, suddenly, there was a shift in the wind. The flames swept in the opposite direction. For the moment, at least, the homes were safe.

When the winds finally cooled, Australia had lost 800,000 hectares (2 million acres) of forest. Three firefighters and one citizen were dead, and about 800 people had lost their homes. In the summers since 1994, Australia's bushfires have threatened more towns and suburbs, leading people to worry and wonder. Do they need to change the way they manage the forests? Or is the world itself growing hotter?

Firefighters in British Columbia and California face similar questions. In August 2003, more than 1,000 separate fires attacked the dry woodlands of B.C. A massive forest fire near the city of Kelowna sent almost 30,000 residents to emergency shelters and destroyed more than 200 homes.

Meanwhile, the weather in California grew hotter and drier each day. In October 2003, the forests erupted. Flames roared through the mountains outside Los Angeles and San Diego, killing 16 people and destroying 1,600 homes. Tens of thousands of people fled the region's mountain resorts as the flames surged closer. Rising humidity and dropping temperatures finally slowed the blazes in November. However, by then, the California governor estimated the fires had caused $2 billion in damage.

In North America, as in Australia, forest authorities concentrate on preventing such wildfires. Once the flames get out of control, they grow so hot and feed on so much forest fuel that they are almost impossible to stop. Thousands of firefighters can work around the clock with little effect unless cool temperatures or rains arrive. These blazes seem to tell us that despite our firefighting technology—our alarms and hydrants, our sprinklers and engines—we are sometimes helpless to control the power of fire.

What Would You Do?

The smoke alarm blares in the middle of the night. You wake to the acrid smell of smoke. Fire! Your mother or father pounds on your door. Hurry! How will you escape? What will you take with you?

We've all wondered how we would react in such an emergency. At school we've even practiced quickly filing from the classroom and gathering outside. But in the calm of a sunny day rehearsal, it's hard to predict the panic that a real fire would bring.

The following chapters describe some of the most destructive fires in human history, from the Great Chicago Fire to the Kuwait oil well fires. In some tales, heroes rushed to the rescue. At the Chernobyl nuclear power plant, firefighters entered radioactive zones to help prevent more explosions. In New York, quick-thinking university students saved workers from the roof of a burning factory. In other stories, people fled from danger. When a theatre in Vienna, Austria, burst into flames, the crew and fire wardens ran for the escape routes, leaving the audience trapped.

As you read these stories, stop to imagine the blinding flames, the choking smoke, and the waves of heat. What if you had been there? What would you have done?

Fleeing the Flames

Chicago, 1871

JULIA LEMOS PULLED HER SHUTTERS CLOSED on the evening of Sunday, October 8, 1871, fighting against the tug of the wind. It would be a bad night for a fire, she thought. With almost no rain in Chicago for more than three months, flames could spread quickly. She pulled her shawl more closely around her shoulders. Even with the shutters closed, the harsh wind seemed to push its way into her wooden house.

For Julia and the other residents of Chicago that autumn, fires were common occurrences. The sound of sirens and the sight of fire trucks careening around corners were starting to seem ordinary. Wooden buildings and shacks crowded against one another in the poorer neighborhoods, and because of the lack of rain, it took only a small spark to start a major blaze. The city's understaffed fire department was battling about two blazes a day. Just the night before, firefighters had spent much of the evening battling a fire that had burned several blocks.

Julia was also missing her children very much that evening. She worked as an artist in a printing shop and made enough money to support her elderly parents and her five sons and daughters. But because her mother was sick, there was no one to take care of the children while Julia was at work. She had placed four of them in an orphanage, and she paid a weekly fee for the staff there to care for them.

Despite her worries, Julia eventually fell asleep. But at 5 a.m. she woke to a rumbling outside her window. When she threw open her shutters, she found the street full of people. Some were carefully packing their things onto carts. Other were panicking, running through the street with shawls and nightclothes blowing behind them. The sky above the crowd was lit by a strange orange glow.

Julia spotted a neighbor below. "What's going on?" she called.

"The city has been burning all night," the neighbor said. "It's coming to the North Side."

Julia gasped. That meant the fire was approaching her home. She rushed to wake her parents and her baby. Then, leaving them to prepare, she raced through the streets to the orphan asylum to collect her other children. She found the women there panicking and the orphans running wild. The asylum was in such chaos that her children couldn't put on their proper clothes or hats— they rushed away in whatever they could find.

When she got back home, Julia helped her father pack the family's most precious belongings into trunks and load them onto their landlord's wagon. They also loaded their feather mattress and a few treasured pieces of furniture. The landlord set off to a more northern part of the city, hurrying to save his cargo from the blaze, yet Julia and her family waited in the house and hoped that the wind would change and their home would be spared. Meanwhile, the orange glow in the sky grew brighter. They began to smell smoke on the wind.

A Timber Town

A booming metropolis of commercial buildings and bustling piers, international traders and hardworking laborers, Chicago, Illinois, lay at the mouth of the Chicago River, where the waterway emptied into Lake Michigan. In the previous two decades, a canal and four railroads had opened, making the city a center for trade and a supply depot for the West. Cargo ships plied the river, loaded with timber from the sawmills, wheat from the 17 grain elevators, or goods produced by the 1,100 factories. In 1837, Chicago had been home to 5,000 people. By 1870, it was a city of 300,000. Many of the new residents were immigrants, and some were desperately poor. They lived in tents and shacks grouped haphazardly into shantytowns.

The fire that threatened Julia's home early on the morning of October 9 had started the evening before in the O'Leary barn. Catherine O'Leary lived with her husband and her five children in the back of a wooden cottage in the western part of the city. She made her living selling milk to the neighbors. But when Mrs. O'Leary milked her cow on the evening of October 8, she forgot her lantern in the barn. The cow kicked it onto the dry straw, and within seconds the building went up in flames.

By the time the already exhausted firefighters arrived, they were too late. The entire neighborhood was burning within an hour. The streets turned into rivers of fire as the flames tore along the wooden-slatted sidewalks and the wood-paved roadways from home to home, until they reached an industrial district northeast of the O'Leary barn. Soon the timber mills were blazing. Logs and lumber acted like fuel in an enormous, city-sized furnace. Wooden warehouses were transformed into towers of smoke, and coal stockpiles glowed crimson red, ready to feed the inferno.

Before firefighters could even begin building firebreaks, the flames jumped the south branch of the Chicago River and attacked the Southwest Side. About midnight, the blaze moved north to Conley's Patch, a shantytown of mainly Irish immigrants. There it tore through tents and shacks so quickly that many people were trapped. They woke to find their neighbors' homes on fire and their own rooms filled with smoke.

The narrow lanes between the dwellings were soon choked with fleeing residents. People pushed against one another to escape. Men yelled that the end of the world had arrived. Children screamed as they lost their parents in the crowd. Husbands and wives were separated. Soot and smoke blackened people's faces until they couldn't recognize each other in the crush, and those who fell were trampled by the mob. Many of the sick or old were forgotten in their shacks. As smoke filled the alleyways and the heat grew more intense, hundreds of people were killed. The blaze grew so hot that firefighters were forced to retreat step by step from the raging furnace, unable to help those trapped within.

In the midst of the chaos, the flames reached the municipal gasworks. A massive holding tank exploded, and streetlights across the city were instantly extinguished. As the panicking crowds were plunged into darkness, familiar streets became confusing passageways. The glow of the flames offered the only light.

An hour later, another branch of the fire approached the courthouse. From within the high bell tower, a watchman saw the flames coming. They were several blocks, then one block, then only a few buildings away. Suddenly a gust of wind stirred the fire. The tower itself was ablaze! The watchman rushed to the stairs to discover that the flames were already inside. With no time to run down, he perched on the banister and slid his way to safety. Once on the ground, he and the guards rushed to the basement to release the prisoners there. Everyone scrambled to safety. They had barely escaped when the massive courthouse bell crashed down from the tower to the levels below.

Firefighters seemed to be having no effect on the blaze, partly because their forces had been split in two. Half were battling the fire on the South Side, where the flames were consuming offices, the newspaper building, and the homes of the wealthy. The rest of the firefighters concentrated on the North Side, where they battled unsuccessfully to save the homes of German and Scandinavian immigrants. Then firefighters found themselves powerless to stop the flames from reaching the city's water pumping station. When the roof caved in at 3:30 a.m., water pumps across Chicago ran dry. There was nothing more the fire department could do. The exhausted men lay down their hoses and focused instead on finding safety for themselves and the people still rushing through the streets.

Daylight on October 9 brought no relief. The city's largest department store burned, along with the First National Bank Building, the Custom House, and the Post Office. Below the Custom House and Post Office, large brick vaults had been lined with eight centimeters (three inches) of metal. More than a million dollars was stored inside. Officials believed that the safes were fireproof, but the buildings above grew so hot that the bills in the safes burst into flames, incinerated as if they were cooked in an oven.

Some people who lived in untouched districts reported to work in the buildings downtown. Made of bricks instead of timber, these buildings were supposedly fireproof, but workers watched in disbelief as the heat melted the mortar between the bricks. Snatching what they could, the workers joined the crowds fleeing for safety.

It was a terrifying race to the edge of the city. People had abandoned carts and pieces of furniture along the route, and the piles of belongings blazed. Along with the homes and businesses, the wooden streets and sidewalks were burning and some of the bridges were also in flames. When oil that had been spilled on the surface of the river caught on fire, the river itself seemed to burn.

On the North Side, advancing flames herded many people closer and closer to the edge of Lake Michigan. Eventually, there was nowhere else to run. Surrounded by flames on three sides and with the Great Lake on the other, families were forced into the water. A few were rescued by boats, but many of the vessels usually anchored at the shore were already burning, set on fire by flying debris. Dozens of people were left trapped in the frigid water for hours.

Escape!

In the mid-morning of October 9, Julia Lemos and her family were still waiting, hoping that the fire would spare their home. Julia was standing in the doorway holding her baby when a woman rushed past and yelled to her.

"Ain't you going to save those children?" she called.

Julia was startled into action. She and her mother packed a few supplies, then the family set off toward the northern border of the city, navigating through the smoke. Ash drifted from the sky into their hair and onto their clothes. At one point the flames were visible, moving and flickering in the distance almost as if they were alive. Finally, at the edge of the prairie, Julia's family managed to find their landlord and the wagon full of their possessions. She put the children to sleep on the feather mattress and watched as the fire grew closer and closer.

From their new vantage point, great tongues of flame seemed to sweep steadily through the city toward them. Before long, the fire had raced through the rest of Julia's neighborhood, and red-hot embers and bits of wood were falling on her and her family again. This time Julia and the people around her were forced to leave their possessions. Looking back, she saw her belongings catch fire, glowing like a bonfire against the prairie grass. She could feel the fire's heat on her back as she herded her family along, desperate for safety and rest.

They ran until they were exhausted. As the sky darkened and rain began to fall, they found an abandoned farmer's shed and took shelter. In the morning, Julia's father trekked back to the edge of the city. He found the family's possessions—a blackened pile of rubble—smoldering at the edge of the flattened, smoking city.

Smoldering City

By the time rain finally extinguished the Great Chicago Fire on the evening of October 9, a total of 860 hectares (2,100 acres) lay in ruins. The fire had devoured 1,600 stores, plus 60 factories in the business district alone. Three hundred people had died, many of them victims of the uncontrollable blaze in the Irish shantytown. Of the 300,000 people who lived in Chicago, 100,000 lost their homes.

On the morning of October 10, Charles C.P. Holden, a city councilor, walked to the north border of the city, where the fire had swept to the edge of the prairie. There, Holden found thousands of refugees—camped, huddled on blankets, or sitting on the grass, for as far as he could see. Many sat silent, in shock. Others wandered through the crowd, calling for lost family members.

Behind Holden, the city lay like a charred field. A few blackened beams still stood among piles of brick. Here and there, a building frame leaned precariously, its insides gutted and gone. Some pieces of walls stood undamaged, out of place amidst the destruction.

It was hard to know where to begin a cleanup effort, but Holden and other city leaders set to work. In the undamaged parts of the city, schools and churches were converted into temporary shelters. Help poured in from surrounding towns and cities, and carpenters were soon at work building 200 temporary shacks. Trains, moving on the unburned sections of track, offered free passes—and some people joined their relatives in nearby towns. Others traveled to cities such as New York or Des Moines, Iowa, where work was available.

Julia and her family spent several days in a church shelter. Around them, the world seemed to be growing more dangerous. A mob hanged a man accused of helping to spread the fire. Looters wandered the rubble, and Julia even saw a man try to steal a watch from someone sleeping inside the church.

The Chicago printing companies where Julia might have found work had all been destroyed. Although relief agencies were bringing food daily, Julia knew that she couldn't depend on their help forever. With her home and workplace destroyed and her family devastated, she became one of the refugees. She moved her family to New York and worked there for several years before returning to Chicago.

A New City

Chicago rebuilt quickly, intent on proving itself to be one of the most advanced cities in the nation. A year after the fire, a boom in projects allowed architects to imagine new designs—buildings rose stronger and higher than ever before.

Inventors also examined fire-fighting equipment, creating several improvements. The famous firehouse pole that allowed firefighters to slide from their barracks to the garage was invented in the city in 1874. And Chicago was the birthplace of the firefighter's helmet, similar to the ones still worn around the world today.

Struggle for the Jungle

Indonesia, 1997

FLYING JUST ABOVE INDONESIA'S JUNGLE CANOPY in the fall of 1997, the American pilot dripped with sweat. The weather was hot and humid—he constantly felt like he needed a shower. His clothes were sticking to his skin. Around the plane, the air was a thick haze of smoke from the hundreds of forest fires burning throughout the country.

This was far different from any wildfire the pilot had encountered in his home state of Wyoming. There, he'd seen brushfires threaten the prairie or flames run from treetop to treetop. Here, the ground itself was smoldering. The jungles were interspersed with peat bogs, where plant material decaying over hundreds of years had formed a layer of highly flammable fuel. Within the peat, the fire seemed to burn endlessly, attacking surrounding trees from the ground up. As the pilot navigated carefully, only 46 meters (150 feet) above the ground, he saw a massive hardwood tree topple. It had been burned away at the roots by the simmering flames.

Resisting the urge to cough, the pilot squinted ahead through the smoke until he found his target. Then he released the plane's 11,000-liter (3,000-gallon) cargo of fire retardant. Pouring out behind him like thousands of bottles worth of Gatorade, the liquid painted a bright orange ribbon in the yellow-gray sky.

He banked the plane to return for another load. For as far as he could see, flames and smoke rose from the earth. In fact, the fire seemed to stretch right over the horizon.

Fire Starters

On the island of Borneo, reporter Sander Thoenes gazed at the land beside a winding dirt road. This was a development known as the One Million Hectare Project, where the government had granted plots of land to poor farmers. The laborers were supposed to grow rice, but found it almost impossible to grow anything in the poor soil, even if they managed to clear the land.

From the road, Thoenes could smell ash and smoke and see shriveled grasses and blackened tree trunks, damage left by the fires that had passed through the month before. After weeks of struggle, firefighters had managed to dig veins of smoldering peat out of the earth and hack down swaths of forest to create firebreaks that the flames couldn't cross. With the blaze finally dampened, they had departed to fight similar battles nearby.

As Thoenes surveyed the devastation, he saw two farmers gathering branches and sticks. At first he assumed they were attempting a cleanup operation. He watched them pile their collections at the edge of the charred area, then scrounge for more until the piles reached waist high. Suddenly he gasped. The farmers had just set fire to the stacks of wood! With the land so dry, and no rain expected, the bonfires would be enough to spark the surrounding trees and reignite the forest fire. When firefighters were trying so hard to extinguish wildfires, how could these men set more of the bog ablaze?

When Thoenes told others what he had seen, he learned that local farmers saw no other choice. They were desperate to feed their families, and that meant clearing the land to plant crops. And the only way they knew to clear the land was to set it on fire.

Traditional Indonesian farmers who grew rice and other food crops weren't the only ones setting fires in 1997. In the quickly developing country, there were also companies that needed open fields for coffee, cocoa, palm oil, and rubber plantations. The fastest and easiest way to clear the jungle was to set it on fire, so plantation owners and individual farmers set land ablaze each summer. The Indonesian government had officially made burning the jungle illegal in 1994, but the laws were never enforced. Many of the plantation owners had friends or relatives in high government positions, and they used their connections to protect themselves from prosecution. Usually the monsoon rains of fall would put out the flames, leaving the land ready for agricultural use in following years.

In 1997, farmers and plantation owners set 200 fires on the island of Sumatra and 650 more in the province of Kalimantan on Borneo. Then the monsoon rains didn't arrive. A weather pattern known as El Niño, which occasionally changes the currents of the Pacific Ocean and the air currents above, brought the worst drought in 50 years to the islands of Indonesia. The fires that had started in June continued to burn and spread, and the entire country was soon blanketed in a choking haze.

In some parts of Indonesia's jungle, fire raged through stands of mangrove trees and nipa palms, burning rainforest vines, rare orchids, lilies, and carnivorous plants. Tropical hardwoods blazed like torches, and banks of flame consumed flowering shrubs and canopy leaves within minutes.

Other parts of the jungle were too damp and humid to burn with such force. Instead, the fires burrowed underground, deep into the peat. Scientists used satellites to track the fires' progress and identified "hot zones" on their maps. When hot zones disappeared, the scientists would consider the fires in those areas extinguished. But then the hot zones would reappear days or weeks later, the fire having smoldered in the peat, slowly burning through tree roots and undergrowth all that time. This type of fire seemed much less dramatic that the roaring jungle blazes, but the effects on plants and animals were just as deadly.

It didn't take long for surrounding nations to notice the effects of the fires. First the clouds turned a dingy yellow, then the sun disappeared altogether. By the end of September, the sky was gray over Malaysia, Singapore, the Philippines, Brunei, and Thailand. In Kuala Lumpur, Malaysia, businesspeople wore surgical masks as they walked the smog-filled streets.

The American Environmental Protection Agency uses a system for rating air pollution, the Pollution Standard Index. When the index reaches over 100 in American cities, people are warned that their asthma or breathing problems may become worse. In Kuala Lumpur in September 1997, the index was regularly at 300, 400, or even 500. In the Malaysian state of Sarawak, it reached 839. The Malaysian government declared a state of emergency. Schools and businesses were closed when people could no longer see more than an arm's length in front of them.

On September 26, a Garuda Airlines flight traveling from Indonesia's capital city, Jakarta, to Sumatra crashed only a short distance from its destination. Flying low through the smoke and haze, the plane hit a tree and fell into a ravine. The 234 passengers and crew members died instantly. Many blamed the crash on poor visibility due to the smoke, and other airports in the region closed completely.

In the Strait of Malacca between Indonesia and Malaysia, an Indian ship and a Caribbean ship collided, unable to see each other through the thick smog. Twenty-eight people were killed.

Fighting the Unknown

On the island of Sumatra, a father sat huddled in grief, mourning the loss of his six-month-old son.

"He was fine when he was born," the father told an American reporter. "Then he grew more and more sick. I don't understand it."

Two daughters sat farther inside the small hut, coughing on the smoke from the surrounding fires. Though the hut was surrounded by leafy trees, the air felt heavy. Ashes coated the ground outside.

"Do you think the smoke could have killed him?" the reporter wondered.

The father shrugged, too grief-stricken to care. Besides, there were no doctors in this remote area of the country, so there was no way to know for sure why the little boy had died.

In villages throughout Indonesia, poor workers were facing similar questions. Doctors and scientists around the world were raising alarms, warning about deadly cases of asthma and lung infections caused by the smoke. But the news never reached many rural Indonesian villagers. With no access to doctors or hospitals, they did not know how dangerous the smoke could be. In some villages, clouds of charred leaves and ashes swirled constantly from the sky.

A teacher in a Sumatra elementary school told a reporter that there were no health problems associated with the fires. But, she added, the teachers could no longer use the blackboards. The smoke was so thick inside the classrooms that students couldn't see to the front of the rooms.

By early October, 35,000 Indonesian people and 15,000 Malaysians had been treated for respiratory disorders. Western doctors said that just breathing the air was equal to smoking 80 cigarettes each day.

A Blind Eye

As neighboring nations complained about the haze disrupting trade and tourism, the Indonesian president publicly apologized. Other countries worried that he was saying the right things but doing nothing about the problem.

By early October, 400,000 hectares (almost a million acres) were burning—an area larger than Rhode Island and almost half the size of the island of Hawaii. Indonesia finally sent firefighters, but they were armed only with shovels and sticks. Even when they managed to smother the flames, the local farmers often reignited them.

The international community began to take action. Malaysia sent 1,200 firefighters, and Canada, France, and Australia pledged equipment and experts. Japan sent specialized water cannons, and the United States called on its National Guard and sent 90 firefighters, six Forest Service experts, and three planes. These forces descended on Indonesia determined to quickly conquer the flames.

Fleeing the Jungle

An American environmentalist walked through a village near his camp. The smoke was so thick he could taste it in his throat. For weeks, everything had smelled like ash to him. The fires were several miles away from the camp, but he had occasionally seen the orange glow of flames late at night.

Suddenly he stopped in his tracks. There! By the side of a shack, a woven cage held a tiny orangutan, not nearly old enough to be taken from its mother. The environmentalist shook his head—this was probably one of many animals caught fleeing the burning jungle. In the past weeks, tigers, elephants, Malaysian sun bears, and flying foxes had wandered onto plantations to scavenge for food. They were often found and killed by plantation workers.

Hundreds of endangered orangutans had also been killed. Villagers shot both the large, aggressive males and the smaller female orangutans, collecting their young to keep as pets or to sell. Though the trade in baby orangutans was illegal, it was extremely profitable.

The environmentalist hesitated. According to the rules of his group, he was never, ever allowed to buy an animal from the villages—it might encourage people to capture more of the animals. But how could he leave this tiny animal here? Quickly he bought the creature and carried it back to the camp. Other workers there were already preparing to spend years working with rescued animals and releasing them back into their natural habitats.

A Never-Ending Battle

As the weather in Southeast Asia finally cooled in late October and early November, firefighters began to think the worst was over. A thousand Malaysian firefighters declared victory over the blazes in Sumatra. The flames were extinguished—the undergrowth was no longer burning and the land was an ugly but cool expanse of charcoal and ash. The firefighters returned home.

Only four days later, they were called back—the fires had reappeared. At first there were just a few hot spots, then the jungle was bursting into flames just as fierce as those that had swept through the trees weeks before. It seemed the peat bogs had fooled both the firefighters and the satellite systems again. The fire had simply gone underground during the firefighters' battle, burrowing beneath the bogs while workers poured water over the jungle's surface.

At the University of Palangkaraya, students lost their faith in the methods of the firefighters. They decided to battle the blazes themselves and rigged an electric pump in a water well. They then dug a long ditch along the upper edge of the school grounds. By filling the ditch, then overflowing it, they sent a slow but steady blanket of water over the fields. The peat fires were beaten, at least in one small area.

Breathing Again

People across Southeast Asia and around the world breathed sighs of relief—and gulps of clean air—as the monsoon rains finally arrived in December. But the Indonesian government continued to downplay the extent of the damage. Hardly anyone had died, officials insisted, and only 170,000 hectares (420,000 acres) of jungle had been destroyed—an area smaller than some of the country's national parks. Relief organizations believed the fires destroyed ten times that area of jungle. Experts also believed that 30 million people in Asia suffered health problems because of the smoke.

All of this meant little to the farmers and plantation owners of Sumatra and Borneo. As the dry season of 1998 wore on, they once again set fire to the jungle. More hardwoods blazed, more peat bogs burned, and more ash rained down on the local villages. The monsoons arrived on time that year and extinguished the flames, but the fires have been relit on the islands of Indonesia every summer since. Each year there is less jungle to burn.

A Nuclear Nightmare

Chernobyl, 1986

THE EIGHT-YEAR-OLD BOY woke with a start. He heard a boom, like amplified thunder. It was 1:24 a.m. on April 26, 1986. His mother poked her head into his room. She said that a plane had probably crossed the sound barrier above their house. It was nothing, she said. Go back to sleep.

When he woke up for school later that morning, he found the street outside coated with green foam. He recognized the slimy substance. It was used to wash the streets when the nearby Chernobyl nuclear power station released any radioactive material. It wasn't supposed to be dangerous.

His mother walked him to school, chastizing him for kicking and sliding in the foam along the way. It was bright and sunny outside, and it seemed like any other day. But almost as soon as he sat down, he could tell something strange was going on. His teacher, usually strict and efficient, seemed distracted. She kept peering out the classroom windows and scurrying into the hall for whispered conferences with other teachers. She told the students to read quietly to themselves.

Finally, just after lunch, a doctor made rounds through the classrooms to give each child an iodine pill. The boy didn't understand why, but he took the pill without complaint. The teacher had just announced that school was canceled for the afternoon and told her students to go straight home.

They were soon pouring out into the sunshine. The boy intended to go directly home, but then he was distracted by an ambulance rushing by, sirens wailing. Next he noticed a police car circling the neighborhood. What was going on? Was there a fugitive on the loose? An escaped criminal? He heard the *chop-chop-chop* of an approaching helicopter. It landed in a nearby parking lot, sending up a whoosh of dust and pebbles in the wind from its blades. He was too far away to hear their voices, but he saw medics scramble out of the chopper. They unloaded two men on stretchers. Another man, dressed like a firefighter, began throwing up on the pavement. Another ambulance screeched by and pulled up beside the injured men.

What was going on? The boy turned toward home again, walking faster, then running. This was his dad's day off—maybe he would know what was happening.

Problems at the Plant

The boy and his family lived in Pripyat, a city built near the U.S.S.R.'s Chernobyl nuclear power plant. Many of the 45,000 people who lived there worked in the plant or provided services for the workers. The plant was named for the smaller town of Chernobyl, about 15 kilometers (9 miles) away.

The problems at the power plant had begun the day before, on April 25. The managers had decided to test whether one of the nuclear reactors—reactor No. 4—could still function in a power outage. During the switch between normal power and backup power during an outage, could they still continuously fuel the pumps that sent cooling water into the nuclear reactor's core? If the pumps failed, the reactor could overheat and cause a nuclear disaster. Fortunately, similar experiments at other nuclear facilities had gone well and the power had remained stable.

At 2 p.m. on April 25, workers shut off the emergency core cooling system to prepare for the experiment. This was against regulations, but the workers weren't worried. The reactor had never had any serious problems. Unfortunately, the experiment was delayed, because the nearby city of Kiev needed extra power for the evening. City officials asked the plant managers to keep the power plant operating at full capacity until later that night. The managers agreed, but no one reconnected the emergency cooling system.

Just after midnight, the experiment resumed. Technicians had some trouble stabilizing the power being produced by the reactor, but they eventually succeeded and the tests continued. What they didn't realize was that the water used for cooling the reactor was growing hotter and hotter. Because warm water didn't work as efficiently as cold water, the water pumps had to send twice as much water through the plant. The pumps themselves began to overheat.

At 1:22 a.m. a plant operator received a computer printout with details about how the plant machinery was operating. He saw that the heat levels were unsafe. According to the rules, he was supposed to immediately shut down the reactor. Instead, he allowed the experiment to continue.

Just one minute later, at 1:23, the power levels inside the reactor began to rise dramatically. The core grew hot. The foreman ordered an emergency shutdown, but it was too late. He could not cut the power quickly enough, and the entire reactor overheated. Steam and hydrogen gas accumulated inside.

At 1:24, the reactor exploded.

Hydrogen trapped inside the reactor ignited, and a globe of burning gas shot into the air. Flaming wood and metal rained down on the surrounding buildings, setting their tar roofs on fire.

The roof of the reactor itself weighed 1,000 tonnes (1,100 tons). It was lifted into the air as if it were a piece of paper, then slammed back down slightly crooked, leaving the sides of the reactor gushing radiation onto the power plant grounds. Some of the reactor's protective metal coating melted in the heat. Forty-nine tonnes (54 tons) of nuclear fuel exploded into the sky above Pripyat, rising up to a kilometer (about half a mile) into the atmosphere. Sixty-eight tonnes (75 tons) enveloped the reactor itself, contaminating equipment and the surrounding buildings. The water used to cool the core was now highly radioactive and it was pouring from the torn sides of the building.

What had been a quiet, dark night a moment before erupted like a volcano. Along with the reactor itself, now transformed into a searing-hot oven, 30 surrounding buildings caught fire. Each structure seemed crowned by flames as the fire raced across the tar roofs, sending crimson streaks into the sky. As emergency alarms wailed and workers shouted to one another, the walls of the buildings began to crack and pop. The scene looked like something from a firefighter's nightmare.

Into the Danger

Lieutenant Vladimir Pravid was chief of the emergency crew on duty at reactor No. 4 that night. He felt the blast shake his building and he scrambled outside to his truck in time to see flames shooting from the roof of the reactor. He immediately radioed a Code 3 alarm. Code 3 was used only in dire emergencies. He knew that firefighters from Pripyat, Chernobyl, and even Kiev would soon be rushing to help. Off-duty workers would be donning their uniforms and converging on the fire stations.

Meanwhile, the fire was spreading by the second. He saw the roof of a building burst into flames as a chunk of flaming metal fell onto the tar. Soon the structures behind and in front of him were burning. His crew's fire engines raced into positions around the reactor. Men leapt to connect fire hoses to water hookups while others scrambled onto the roofs for a better perspective on the flames. Still others scouted the surrounding buildings. Was it safe to enter? Were survivors trapped inside?

Major Leonid Telyatnikov and his team of Pripyat firefighters also heard the blast. They were rushing to their trucks even before they received the Code 3 alarm. When they arrived at the scene they found the normally calm plant transformed into a battle-ground. Firefighters had dragged their hoses up ladders and onto the roofs. They stood against the black night sky, their faces lit by the glow of the flames. The roar of the fires and the howls of approaching fire engines combined in a deafening clamor. Some men yelled to one another. A few plant workers were led scream-ing from the site, their skin blistered and charred from the flames. Mostly, though, the firefighters battled in determined silence, focused only on preventing the spread of the flames. Reactors 1, 2, and 3 lay dangerously close.

The firefighters were not equipped with specialized gear to protect them against either the radiation or the intense heat. On the roof of the control room, two men battled the blaze with such intense concentration that they suffered burns on their arms and legs. By the time other firefighters arrived, the first two were staggering with exhaustion. The newly arrived crew members led them to the ladder, sent them toward the waiting ambu-lances, then took their places against the flames. By the time the fire was extinguished, this second crew also had to be evacuated by ambulance.

By 2:30 a.m., barely an hour after the blast, the firefighters had succeeded. The reactor core continued to smolder, but the surrounding fires had been put out. The plant lay dark again. Paramedics distributed iodine tablets to the men to help protect them against radiation poisoning. Ambulances took the last of the injured to the local hospital.

When the reactor had exploded, there were 176 staff members on duty and 268 construction workers building two new reactors nearby. Of these workers, 132 were hospitalized with injuries and one died of his burns. One plant worker disappeared and is believed to be buried in the wreckage of the reactor core.

Six firefighters died that night, and four died later of their injuries. In the crowded hospitals, more than 200 people struggled to survive extreme radiation poisoning. Many had begun vomiting only minutes after exposure. Now they had dangerously high temperatures and diarrhea. As the mucous membranes in their noses and mouths swelled, they found it painful to swallow or breathe. Some hallucinated in the grip of their fevers, calling out to friends or family members who weren't really there.

In the aftermath of the explosion, people wondered whether the firefighters had understood the dangers they faced. Did they realize the reactor was pouring out radiation? Were they aware the power plant grounds were as toxic as an atomic bomb site?

They may not have known the extent of the disaster, but the stories firefighters told from their hospital beds proved the men knew they were taking great risks. They had watched ambulances whisking away other firefighters who had suffered radiation poisoning and burns. Still, replacement crew members had poured in. Some had seen the molten core of the reactor itself, yet they had stayed to fight the blaze. If they had chosen to leave the scene, the other reactors might have soon caught fire. The radiation that attacked them might have spread through more of the nation. The firefighters' selfless dedication helped to contain the disaster.

Evacuation

After the eight-year-old boy arrived home from school on the afternoon of April 26, his parents kept him indoors. They still weren't sure what had happened. The radio called it a minor incident at the power plant. Then, on the afternoon of April 27, the radio announced an evacuation. The boy's mother tore through the house, packing food, clothes, and important papers. She didn't know where they would be going. The radio said to pack for several days. His father phoned some friends who worked at the plant. They said to prepare to leave forever.

At 3 p.m., the boy stood with his mother on the sidewalk and watched people file onto a long row of buses. His father had been called in to help out at the plant. He would follow later, he had said.

A few people waiting for the buses threw up on the sidewalk, overcome by the effects of the radiation in the air, on the streets, and even on their clothes. Many were dizzy and some couldn't stand. Still, no one was left behind. In less than three hours, the entire city was empty. The bus that carried the boy and his mother was part of a caravan of more than 1,200 vehicles winding slowly away from the power plant.

The Response

When the government of the U.S.S.R. evacuated Pripyat on April 27, many residents were taken to Kiev, just over an hour away. By May 2, everyone living within a 10-kilometer (6-mile) radius of Chernobyl was evacuated. Everyone inside a 30-kilometer (19-mile) radius was moved by May 5. In total, 116,000 people were forced to leave their homes. They were taken in by residents of Kiev and surrounding villages, and temporary housing was eventually found. The government offered compensation and found workers new jobs, often at their previous wages.

Despite the relatively quick response inside the country, the government tried to hide the disaster from the outside world. It wasn't until April 28 that the international community realized something was wrong. Swedish scientists detected radioactive fallout over Scandinavia, and it seemed to be blowing in from the U.S.S.R.

Meanwhile, the core of the Chernobyl reactor was still smoldering at 1500°C (2800°F), hot enough to melt blocks of gold, silver, or iron as if they were ice cubes. The core seethed like a red-hot coal. Water turned to steam before it even neared the fire.

The only safe way to approach the furnace was from the air. After being specially shielded with lead, helicopters began dumping load after load of lead, boron, sand, and clay onto the core, hoping to seal it. Workers also began to build a concrete layer below the reactor's foundations, to help keep the radiation from leaching into the groundwater.

Eventually, engineers, scientists, and construction workers collaborated to create an enormous steel frame over the radioactive core. Each steel beam weighed 150 tonnes (165 tons), or as much as five full cement trucks. Once the frame was erected, workers began to fill it with cement, pouring about 500,000 cubic meters (650,000 cubic yards) into the depths and creating a looming sarcophagus—a massive concrete structure—over the site. A cooling system inserted within the sarcophagus removed excess heat from the still smoldering core. Three months after the accident, the combination of shielding succeeded in blocking the radiation.

The U.S.S.R. separated into Russia and a number of smaller nations in 1991. Chernobyl now lies within Belarus. Because of the accident, radioactive components such as plutonium have entered the soil. Not only does the radiation make trees and animals sick, it can also cause cancer in people who drink the water or eat plants grown there. About one-fifth of the region's farmland is completely unusable and may remain contaminated for centuries.

Those who were affected by the blast and people who live in contaminated areas continue to suffer the effects of radiation. The birth rate has dropped. The death rate is high among the workers who helped with the cleanup after the accident. Thousands of children in the region have been diagnosed with cancer.

On the Other Side of the World

A supervisor walks quickly through the hallways of a nuclear power plant in North Carolina, checking that the instruments are working properly and the technicians are following proper procedures.

Occasionally, he thinks about the disaster on the other side of the globe, in Chernobyl, years before. So many people dead, and so much land poisoned. But the supervisor quickly reassures himself—power plants are much safer now. Each safety system has a backup, to minimize the risk of human error. Even if he shut down an emergency system like they did in Chernobyl, the reactor would still be safe. This plant has also been designed with something called a negative void coefficient. If the water used to cool the reactor begins to grow warmer, the reactor automatically slows down.

Besides, the technician tells himself, this is only one of 330 reactors producing power in 31 countries. Another 284 reactors are used for research. Belgium, Bulgaria, France, Hungary, Japan, Lithuania, Slovakia, Slovenia, South Korea, Switzerland, and

Ukraine all rely on nuclear reactors for more than one-third of their power, and there hasn't been a major disaster since Chernobyl.

Still, the supervisor continues his rounds, checking each instrument carefully. Sometimes he feels as if the nuclear core is a crater of bubbling lava and he is in charge of preventing an eruption.

Factory in Flames

New York, 1911

EIGHTEEN-YEAR-OLD ROSE ROSENFELD took a break to stretch her shoulders and back. She'd started her shift at the Triangle Shirtwaist Factory at 7 a.m. on the morning of Saturday, March 25, 1911. It was now 4:30 p.m.—almost time to quit.

From her seat in front of a sewing machine, she gazed around the room. The Triangle Shirtwaist Factory occupied the top three floors of the 10-story Asch Building at Greene Street and Washington Place in New York City's Greenwich Village. A group of tailors—mostly men—worked on the eighth floor. Rose worked on the ninth with the sewing machine operators. The executives' offices were on the top level. The company employed 500 young women and 100 men to make shirtwaists, a type of women's blouse.

Most of the workers were recent immigrants from Italy, Russia, Hungary, or Germany. They spoke little English and were happy to have jobs. Some of them supported entire families on their starting wages of $1.50 a week. They worked 14-hour days from Monday to Friday, arriving at 7 a.m. and staying until 9 p.m. On Saturday they got off early, at 5 p.m. Sunday was supposed to be a day off, but it wasn't. A sign near the elevator read, "If you don't come in on Sunday, no need to come in on Monday." The owners of the company, Max Blanck and Isaac Harris, knew that there were thousands of immigrants looking for work in the city.

If employees weren't willing to put in long hours, they could easily be replaced. Besides, hundreds of other factories throughout the city demanded the same long hours.

Rose sighed. No matter how grateful she was to have a job, she still hated staring at shirtwaists all day. Her job was to operate a machine that pressed buttons onto the blouses. Beside her, a row of women sat elbow to elbow, some of them as young as 16. They were lined up at a table that stretched for 23 meters (75 feet) down the length of the room. Another row of workers sat right behind Rose, and more rows sat in front of her. Above their heads, clothesline-like wires carried partially finished shirtwaists to and fro between the workers. As they trimmed and cut, the seamstresses threw the remnants to the floor, until the space between the chairs was filled with bits of cloth. In the corners, some piles of remnants reached waist high. Rose knew it was the same on the floor below.

She bent back to her work, deciding to finish one more piece before the bell rang to signal the end of the shift. Suddenly, a flash of orange in the corner caught her eye. Fire!

It had traveled up the stairs and through the vents from the eighth floor, where someone had thrown a lit cigarette into a pile of fabric. Some of the men and women nearby jumped up to fill water buckets. They tried to douse the blaze, but the scraps of cloth carpeting the floor made this impossible. The fire simply traveled from one pile to another. Within minutes, the lines of cloth overhead were on fire, and flaming bits of fabric were falling on the heads of the would-be firefighters. They gave up their efforts and looked for an escape route.

Rose was already heading toward the main stairway, but she soon remembered such a route was useless. To prevent workers from leaving early or stealing cloth, the owners kept a gate across the stairs. Even though the shift would have ended in only 20 more minutes, the gate was still locked. There was a second stairway, a small fire escape, but too many people rushed toward it at the same time. A few women fell in the crowded entrance. More fell on top of them and soon the doorway was blocked.

Rose joined the rush toward the freight elevators. To get there, she had to scramble over or under the long tables, past overturned chairs, and around more and more burning piles of cloth. The smoke was already stinging her eyes and making her cough. Around her, women were screaming in Italian and Yiddish. She couldn't tell what they were saying, but she could guess. When she finally got close to the elevators, she found one car jammed with girls and the doors closing. The other elevator was out of order.

Women clawed past Rose, reaching for the windows. They screamed to people on the streets, nine stories below: "Call the fire department!"

Rose was jostled, almost knocked over by workers pushing this way and that. Suddenly, she saw someone's skirt catch fire. The flames seemed to form a second layer of clothing, climbing the woman's form as she screamed. A few people tried to help, but they were in danger themselves. Women pressed even closer to the windows.

Rose struggled to stay calm. There had to be a way out. The elevators were jammed, the stairs were blocked. Where else could she go? Then she thought of the owners, Mr. Blanck and Mr. Harris. They would have been in their offices on the 10th floor when the fire broke out, but she hadn't seen them come downstairs. How would they get out? Pulling a piece of cloth over her mouth, Rose fought her way to the stairs leading up. By the time she reached them, she was blinded by the smoke. She felt her way.

At last she made it to the top floor, where the air was slightly clearer. She was just in time to see Mr. Blanck and Mr. Harris disappearing into a freight elevator. She'd never noticed it before— it wasn't one of the ones the women rode in the mornings. She quickly squeezed in after them and soon found herself on the roof of the building.

Falling from the Sky

Outside, a passerby heard screams coming from the Asch Building. He looked up to see a colorful bolt of cloth flying through the air, fabric billowing. Then he realized. It wasn't a bolt of cloth. It was a woman! As she fell to the ground, her skirt billowed out behind her.

Within minutes, a huge crowd was gathered across the street from the building. People stood transfixed as young women inched out onto the ledges above.

"Don't jump!" yelled the people on the ground. The body of the first woman who had leapt lay still on the sidewalk. There was no way to survive such a fall.

But ten stories above, the women seemed to have no choice. Only five minutes after the fire began, flames were already shooting from the windows. The onlookers saw women with their skirts ablaze or their hair on fire. Another jumped, and then another. The crowd at the windows grew thicker, screams echoing between the buildings. Another woman was pushed by the crowd behind her and fell to her death.

At last, the wail of sirens filled the street. It was now 4:50— ten minutes since the first pile of fabric ignited. The onlookers breathed sighs of relief as the fire trucks arrived, only to watch with horror as the bodies of the dead prevented the trucks from driving alongside the building. Finally firefighters found a clear route. They extended their ladders—to the sixth floor. They would reach no higher. With hoses, they could reach the seventh floor, but there seemed to be no way to rescue the women above. Another worker jumped, seeming to aim for the ladder below her. She missed and fell through a glass awning to the ground.

Trapped!

A few women had already burned to death within the building, and many workers were in hysterics. Some pressed themselves against the locked gate in a pointless struggle to escape. Others were trapped between the walls and the flaming piles of debris. Twenty-year-old Cecilia Walker ran with a friend toward the elevator. Cecilia tore open the door. The car wasn't there, but thick cables led down into the darkness. Cecilia's friend gave up and ran toward the windows.

Cecilia grabbed one of the cables, wrapped her legs around it, and began sliding down. She seemed to slide forever. The skin on her hands began to tear, rubbed by the cable. Her legs began to burn. Just as she thought she must be approaching the ground, she fainted. Moments later, she was woken by firefighters, who pulled her out of the elevator shaft along with others who had tried the same route. She was injured, but alive.

Rescue Attempts

From a neighboring building, cloth worker Benjamin Levy heard screaming. Someone said the building next door was on fire. Rushing to the street, Benjamin found the crowd in shock, the fire trucks useless, and the ground littered with bodies. When he saw a group of men trying to stretch a net between them, he rushed to help. They barely had grips on the fabric when the first woman fell toward them. She fell so quickly that she pulled the net from their hands and hit the ground below. She was killed instantly.

The men quickly reassembled. They held the edges more tightly. A second woman streamed toward them, bounced off the stretched netting, and fell onto the sidewalk. A third falling woman ripped the net from their hands again and sent three men toppling backwards. Again and again they tried to hold the fabric. They caught about 15 women in total. Benjamin thought one or two might live.

At one point, Benjamin looked up and saw a young tailor balanced on a window ledge high above him. He was off to the side, out of the range of the men's net. Benjamin watched as the tailor helped a young woman out onto the ledge beside him. She wrapped her arms around his neck. They kissed. Then the tailor grasped her around the waist, held her over the street, and released her. Before she struck the ground, he jumped as well.

Benjamin swiped at a tear. But as he looked at the flames streaming out the window, he understood. The couple had chosen between the fall and the fire. They had said good-bye and jumped together.

Thomas Gregory was an elevator operator in a nearby office building. He was walking home from work when he saw the commotion. Hoping to help, he rushed inside the flaming building. He was able to access an elevator car, and he quickly rode it to the eighth floor. Fifteen women scrambled inside. More clawed at Thomas, trying to get in, but he knew the elevator wouldn't hold them. Forcing the doors closed, he took the 15 lucky workers to ground level. Then he went up again. Thomas made three trips, saving about 45 women, before the elevator was damaged by the flames.

Frederick Newman and Charles P. Kramer were in the middle of a law class in the New York University building next door when someone spotted the smoke. The professor immediately dismissed the class, and most students rushed down to the street. Frederick and Charles stopped a few. They found a ladder and led their group to the roof. There they were able to use the ladder to bridge the gap between the roofs of the two buildings. About a hundred women who had taken refuge on the roof managed to climb to safety.

Freedom

Most of the screaming had stopped. The flames had consumed the piles of fabric, the clotheslines of shirtwaists, the long wooden tables. All the fuel within the brick walls was gone, and the fire subsided. It was only 5 p.m.

Rose Rosenfeld, who had followed the factory owners to the roof, was found there by firefighters and helped to safety. They led her down the stairs of a neighboring building. As she reached each landing, Rose stopped, sat on the stairs, and cried. Her rescuers prodded her along.

One of the first people Rose saw when she reached the ground was her father. He was gazing up at the charred window ledges and Rose could tell by the look on his face that he thought she was dead. She ran toward him, calling his name. He turned, saw her, and fainted.

Rose made a decision as she revived her father and helped him off the pavement. She wasn't going back to work in a factory. Not this factory and not any factory. She was somehow going to go to college.

Grisly Discoveries

When firefighters at last broke into the remains of the Triangle Shirtwaist Factory later that day, they found charred bodies piled in front of the locked stairway gate and the elevators. Few of the women had known about the escape route to the roof.

Newspaper headlines the next day read "141 Men and Girls Die" and "New York Fire Kills 148." The final death toll was 145 of the factory's 600 employees. Most were suffocated by smoke or killed by the flames.

Owners Max Blanck and Isaac Harris escaped without injuries. They were charged with manslaughter and put on trial eight months after the fire. Their defense lawyer said that the two men had subcontracted the labor for the factory, meaning that someone else had hired and supervised the workers. The lawyer claimed Blanck and Harris had not known that the gate in the stairway was locked.

The jury, deciding there was no way to prove that the men had known about the locked gate, found the owners innocent. Blanck and Harris were then able to claim their insurance on the building. They received enough money to start another factory, where they were later fined for locking the workers inside.

The families of 23 dead women pressed civil charges against the factory owners, then settled out of court. Blanck and Harris paid $75 for each woman killed.

The Fight for Change

Without waiting for the courts or the government to act, two labor organizations took action after the factory fire. The International Ladies' Garment Workers' Union had been campaigning for more than a decade for workers' rights. Now they held protest rallies on the streets of New York, demanding better wages for women, better working conditions, and an investigation into the factory fire.

The Women's Trade Union League also mobilized its members. Their vice-president, Rose Schneiderman, gave a speech to many city employers. She insisted that the time for waiting, for cooperating, for negotiating was over.

> I can't talk fellowship . . . Too much blood has been spilled. I know from my experience it is up to the working people to save themselves. The only way they can save themselves is by a strong working-class movement.

The league members collected the testimonies of the factory fire survivors and pressured the government to launch an investigation. In this they were successful. A month after the Triangle Shirtwaist Factory fire, the state governor appointed a commission to investigate factory fires. The commission held statewide hearings for five years and found that more than half of factory fires were preventable. They recommended that employers ban smoking, that they clean up the piles of waste that littered factory floors, and that they provide fireproof receptacles for fabric remnants or other trash. They also suggested that all gas jets on machines be covered by wire mesh or glass globes.

To fight fires that couldn't be prevented, the commission suggested new fire alarm systems. They also explained a recent invention that could revolutionize fire fighting—the sprinkler system—and recommended that factory foremen supervise regular fire drills.

The Triangle Shirtwaist Factory fire also led to more rights for women workers. The public had been shocked by the deaths of so many people, and they supported a shortened workweek, higher wages, and the right of women workers to form unions.

One of the primary campaigners for these improved rights was Rose Rosenfeld. After the fire, she attended college and married businessman Harry Freedman. Throughout her career as a mother and later an accountant, she spoke to labor groups, students, and reporters about the fire and the working conditions of the time. Rose died in 2001 at age 107, the last survivor of the Triangle Shirtwaist Factory fire. Her legacy lives on in the high fire-safety standards at factories and workplaces today.

"It's Going to Blow!"

Halifax, 1917

CAPTAIN AIMÉ LE MEDEC sighed with relief as his ship, the *Mont Blanc*, finally passed the submarine nets and began the journey into Halifax Harbour. It was the morning of December 6, 1917. The *Mont Blanc* was carrying a heavy load of explosives: 2,300 tonnes (2,500 tons) of picric acid and 200 tonnes (220 tons) of TNT. In non-wartime, the TNT alone would have been enough to blast through hundreds of rock walls in a mine or quarry. The ship also carried 10 tonnes (11 tons) of guncotton and 35 tonnes (38 tons) of benzol, a highly flammable chemical stored in barrels on the deck.

World War I had been raging in Europe for three years, and these explosives were desperately needed in France. Troops there were mired in the trenches, fighting back and forth across strips of muddy ground, valiantly trying to hold back the Germans. The French agent in New York ordered so many explosives that he needed more than 20 ships each week to carry the cargo to France.

Like the rest of the French crew aboard the *Mont Blanc*, Captain Medec worried constantly about his cargo. The smallest spark could cause a major explosion, killing them all instantly. They had already removed all oil lanterns and heaters from the ship, and the captain had banned smoking when the vessel left New York.

Captain Medec sighed again when he thought of New York. He had waited in the harbor there, hoping to join a convoy to Europe. Because of the German submarines called U-boats, it was dangerous for ships to sail alone across the Atlantic Ocean. Instead, they traveled in groups, escorted by British warships. Unfortunately, Captain Medec's ship was too old and slow to keep up with the group leaving New York. In fact, the *Mont Blanc* wouldn't be steaming across the Atlantic at all if it weren't for the serious need. The German submarines had sunk so many British ships—more than 1,000 in the past year—that every available vessel had been pressed into service. Every qualified captain was also needed. At 39, Captain Medec had spent most of his life overseeing merchant ships. This was his first journey on the *Mont Blanc*.

As he steamed through the channel into Halifax Harbour, Captain Medec thought some of his worries were over. Every night, Halifax raised massive underwater nets made of steel cables and designed to keep out enemy submarines. When Captain Medec had approached Halifax the night before, he had arrived too late. The nets were already raised. He'd had to anchor his ship alone for the night, an easy target for a submarine. Now that it was morning, he was finally entering protected waters. He was hoping to find a slower convoy leaving from here that could escort him to Europe.

A Wartime Port

Halifax Harbour was one of the largest and busiest ports in the British Empire. Big enough to protect an entire navy fleet, it often sheltered minesweepers and cargo vessels, troopships and battleships. As the main wartime base for Canada, it was the departure point for thousands of North American troops on their way to the front lines in Europe. It was also where shipload after shipload of supplies—weapons, clothing, Prairie grain, and more—was transferred from railcars into waiting cargo holds.

The harbor was shaped like a vase, with a thin neck connecting it to the Atlantic. This neck was called the Narrows. It was a busy thoroughfare, but it wasn't exactly narrow—more than 40 ships could fit side by side within the channel. There was plenty of room to maneuver.

The city itself, home to 50,000 people, stretched along the western shore of the harbor and the Narrows. Dartmouth, a smaller city, lay on the eastern shore. Just before 9 a.m. on December 6, both shores were bustling with activity. Children counted ships on their way to school. Businessmen walked to their offices. Railway workers were already unloading shipments, and factory workers had been toiling for more than an hour.

Marine Maneuvers

On the water, the *Imo* had pulled its anchor and was steaming out of the harbor. Captain Haakon From was impatient—his ship was already 18 hours behind schedule. He had intended to leave the previous afternoon, but loading coal for the voyage had taken too long. By the time the vessel was ready, the submarine nets had already been raised. Captain From had been forced to wait until morning.

The *Imo* was an empty relief ship, on its way to New York to pick up clothing and food for the thousands of Belgian people left destitute by the war. With huge red letters reading "BELGIAN RELIEF" painted on each side of the hull, the *Imo* traveled as a neutral vessel. It didn't need the protection of a convoy.

Usually, ships pass port side to port side. In other words, they stay to their right. When it steamed out of the harbor, the *Imo* was supposed to stay to its right, along the Halifax side, while oncoming ships stayed on the Dartmouth side. As the *Imo* moved through the Narrows, however, it met a cargo ship coming in the opposite direction. Because the cargo ship was moving quickly through the center of the channel, Captain From made a quick decision. He directed his vessel to the left—the Dartmouth side— to pass. Shortly after, he met another ship. Because he was already near the wrong side of the channel, Captain From swung farther to the left and passed the ship safely.

As the *Mont Blanc* steamed toward the harbor, Captain Medec was surprised to see the *Imo* speeding along on the wrong side of the channel. He gave a sharp blast of his ship's horn, telling the *Imo* to move into the proper lane. According to shipping procedures, the first ship to sound its horn is in charge of the encounter. The *Imo* was supposed to follow the *Mont Blanc*'s instructions.

Captain Medec was astonished when the *Imo* gave two short blasts, telling him that it wasn't moving to the correct side of the channel. Instead, the *Imo* was moving even farther left toward Dartmouth.

Captain Medec swerved as close to shore as he could and gave another short blast, again telling the *Imo* to move over. The *Imo* didn't swerve. Instead it gave another two short blasts. It was staying on the Dartmouth side.

Now the ships were bearing down on one another, and there was no time left to maneuver. Frantically, Captain Medec ordered his men to swing around, moving toward the Halifax side.

On the *Imo*, Captain From also saw the danger and yelled to his men, "Full speed astern!" The engines ground into reverse, but it was too late. The bow of the *Imo* sliced into the steel hull of the *Mont Blanc* with a deafening screech. As the ships parted, sparks flew. The *Mont Blanc*, powerless, began to drift toward the

Halifax docks. The *Imo* wasn't badly damaged, but was unable to turn around. It continued toward the Atlantic Ocean.

Within seconds, black smoke was billowing from the deck of the *Mont Blanc*. Captain Medec thought frantically about his explosive cargo. He tried to see whether it was the picric acid or the benzol on fire, but there was too much smoke to tell. Soon, tall tongues of flame sliced out of the smoke toward the sky. The water reflected orange and red as the blaze grew higher and hotter. Soon, sharp blasts echoed across the harbor as the fire reached bullets stored on deck. It was as if guns were firing themselves in random directions.

There was no hope. Captain Medec yelled for his men to abandon ship. They quickly lowered two lifeboats and rowed frantically toward the Dartmouth shore. Along the way, they called warnings to the crews on the surrounding boats. But the sailors from the *Mont Blanc* were French. The other crews either didn't hear or didn't understand their warnings.

Take Cover!

On shore, people had watched in shock as the two ships, with plenty of room to pass each other, had instead steamed directly toward a collision. As fire roared from the deck of the *Mont Blanc*, huge crowds gathered on the docks to watch. They saw a tugboat and two other vessels approach the burning ship, spraying it with water. This did not seem to be helping, and the burning vessel was floating closer and closer to the Halifax piers.

A few experienced sailors on shore recognized the black smoke. They knew that it might mean there were explosives on board. They ran from the docks, yelling warnings along the way.

Many people didn't hear the warnings, but the dispatchers in the rail yard did. William Lovett, the chief clerk, and Vince

Coleman, the train dispatcher, heard the news and rushed to evacuate. Just as he left the building, however, Coleman remembered that a train was approaching the city. He ran back to his desk to telegraph a message: "Munitions ship on fire. Making for Pier 6. Goodbye." The train stopped and 700 people were saved, but Coleman would not escape the building.

Catastrophe

At 9:04 a.m., the *Mont Blanc* exploded. It was the largest manmade explosion in history, only surpassed when the atomic bomb was dropped on Hiroshima during World War II. The Halifax piers were instantly flattened. Timber homes and warehouses were swept away in the shock wave of air. More than 120 hectares (300 acres) of the city were blown away. A gun barrel from the ship flew more than five kilometers (three miles) before it landed. The massive anchor shaft was thrown into the air and landed in the trees three kilometers (two miles) away. Throughout the city, windows shattered and foundations moved. People up and down the coast felt the earth shake and wondered what had happened.

The massive shock wave demolished the harbor and flattened surrounding homes and businesses, instantly killing 1,600 people and injuring 9,000 others. As the air cleared, some people found themselves trapped under fallen timbers or stones and others looked down in shock to find they were missing an arm or a leg. Some people lost consciousness and woke to find themselves blocks away. Liquid tar and ash poured from the sky, and a massive mushroom cloud rose from the harbor.

A tidal wave that followed the explosion swept the *Imo* onto shore near the harbor, killing the captain and six men. The crews of the three rescue boats died instantly, and nine ships sank or were badly damaged. The crew of the *Mont Blanc* had run ashore on the Dartmouth side and thrown themselves to the ground. Only one man was injured and later died from loss of blood.

The wreckage of the *Mont Blanc* set Pier 6 on fire and fire-fighters—those who were still alive—rushed to help. They found the ship blazing like a kiln, so hot that they couldn't turn their faces toward it. The heat singed their hair and skin, tongues of fire exploded higher into the sky, and pieces of blazing metal showered down around them. The men called frantically for more help.

However, little help was available. Other firemen and volunteers were already scrambling to extinguish house fires around the city, caused by burning debris, spilled gas, or escaped hearth fires. Stores of winter coal piled in basements helped fuel more than 30 blazes. With much of the city's fire-fighting equipment destroyed and many people too injured to help, each fire was a new emergency.

For a short while, panic reigned. People screamed in the streets as they rushed home to find their houses gone and their families killed. The wounded and the dead were crushed side by side into the hospitals.

An hour after the explosion, a telegraph message reached the nearby town of Kentville:

Organize a relief train and send word to Wolfville and Windsor to round up all doctors and nurses and Red Cross supplies. No time to explain details, but list of casualties enormous.

Quickly the word spread through the region. An explosion in Halifax. Send help. Within hours, the residents of Boston began to organize a relief train. They rounded up 13 doctors and surgeons, six American Red Cross workers, nurses, and medical supplies. At towns along the way, the train stopped to pick up more volunteers. Montreal, Saint John, and Quebec also sent hospital and supply trains the same day.

When they arrived in Halifax less than 24 hours after the blast, the doctors and nurses found that a blizzard had struck during the night. Survivors who had been trapped under wreckage were now frozen to death. The army was clearing paths for the doctors to access temporary hospitals. Bodies lined the paths, lying in piles three or four deep.

The only benefit of the blizzard was its dampening effect on the city's fires. All the blazes at homes and buildings had been extinguished and Pier 6 had slowly smoldered until little remained of the docks.

Now firefighters and rescuers turned to the debris. One fire captain heard a baby crying from the ruins and found a small girl trapped under a household stove. Twenty-six hours after the blast, he pulled little Annie Liggins from the ashes unharmed. Another infant, two-year-old Joseph Woods, was found alone in the infant ward of the Protestant Orphanage. Twenty-seven children had been killed when the orphanage collapsed and burned, but James somehow escaped the flames and was later adopted.

Aid continued to pour in from across the country and around the world. Australia sent $250,000. Chicago sent $125,000. Ontario sent $100,000. Britain sent a million pounds, the equivalent of $4.8 million at the time.

As the cleanup began, Montreal sent a trainload of horses and mules to help clear the debris. Boy Scout troops began acting as messenger services, compensating for the ruined telephone lines. Women's groups sorted clothing donations and offered free childcare. Sir John Eaton, owner of a large Canadian department store, arrived in town with his private train and distributed free packages of winter clothing.

Halifax survivors would most remember the help that arrived from Boston. After that first train of doctors and nurses, more and more aid arrived. Concert societies held benefit performances, and private citizens donated food and clothing. An army of builders and tradesmen volunteered to travel to Halifax to help construct shelters for the 6,000 people left homeless. In total, the city contributed more than $750,000.

Halifax has never forgotten the help they received during the disaster. Every year since 1920, crews have felled a giant Christmas tree in Nova Scotia. They transport it to Boston as a thank-you for that city's aid in 1917.

The relief committee organized after the disaster remained active until 1976, and the Canadian government still helps to support some surviviors of the explosion. Most were children when the *Mont Blanc* exploded and they are over 90 now. But they can still tell visitors about the deafening blast, the fires, and the families shattered by the explosion.

Disaster in the Desert

Kuwait, 1991

FLAMING OIL ERUPTED FROM THE EARTH, creating a column of fire several stories tall. It blasted into the sky, roaring like a jet engine. The heat seared the surrounding desert. The sand melted into strange glass formations. Black smoke roiled from the tower, creating a kaleidoscope of orange and black that was easily visible from space.

Dwarfed by the flames, looking like ants creeping toward a massive bonfire, three men from a company called Safety Boss huddled behind a shield of corrugated tin. Step by step they approached, constantly pushing the shield ahead of them. They could barely hear over the howl of the fire. They could barely see through the thick smoke. Water cannons positioned behind them blasted the crew with seawater, dousing any possible sparks.

Finally the men were close enough to the burning oil well to direct the rest of the crew. Using bulldozers and a crane, they positioned plastic explosives in a circle around the well. A warning whistle blew, sending everyone running for cover. The earth shook, then there was a brief moment of silence. For a fraction of a second, the explosion had cut off the fire's supply of oxygen. That was enough time for firefighters to blast the well with a torrent of water. Very slowly, the metal of the well and the surrounding sand began to cool. Oil now gushed from the ground, showering the crew and the surrounding earth, but the fire had been extinguished.

Completely coated in brown ooze, the workers looked like bronze statues that had suddenly sprung to life. In the sizzling heat of the desert, the mix of oil and mud that coated their hair, their skin, and their clothes remained extremely dangerous. A single spark could turn a worker or a piece of equipment into a bomb. One of the men, having witnessed too many explosions in the past, worked with his safety helmet duct-taped to his head. If a sudden explosion sent him rocketing backwards, he wanted his head protected.

Although the well had been cooled by the water cannons, the oil continued to gush into the air and the heat remained intense. Kneeling on the ground gave the crew members blisters on their legs. One found that the zipper on his jacket had melted. The water cannons showered the workers with a steady stream of water as well. It was the only way to keep them cool—covered in slime, they were unable to sweat and overheating was a constant danger.

With a few last twists of a giant wrench, the well was closed. The oil was once more locked underground. If these had been the oil fields of Calgary or Houston, where most of these firefighters had been trained, their job would have been over and the danger past. But this was Kuwait in the aftermath of the Gulf War—hundreds of wells were burning. The crew members knew that their success here meant there was a new well, with new dangers, waiting for them the following day.

Oil and Water

Few of the firefighters had ever been to Kuwait before the Gulf War. When they arrived, they found themselves in a desert nation slightly smaller than the state of New Jersey. It lay between Iraq and Saudi Arabia on the Persian Gulf. Their hosts explained that only a very small fraction of the land could be planted with crops. Most of the country's water was supplied by huge coastal desalination plants, which remove the salt from seawater.

But crops weren't Kuwait's main concern. The country thrived on petroleum. Most of the country's 2 million people were supported by the oil industry, an industry that pumped liquid gold from the massive oil reserves that lay beneath the desert sands.

Oil is what had tempted Iraqi leader Saddam Hussein to invade Kuwait on August 2, 1990. Alone, the tiny country of Kuwait wouldn't have lasted long, but the United States and the United Nations rushed to its defense. They began bombing the invading army in January 1991 and by February, the last Iraqi troops were quickly retreating.

But they didn't retreat quietly. Furious about his loss, Saddam Hussein ordered his Republican Forces—the most highly trained branch of his army—to slow their retreat long enough to sabotage Kuwait's oil wells.

Within days, 732 oil wells were blazing and the smoke was so thick in nearby Kuwait City that noon seemed like dusk. Drivers used their headlights even in the middle of the day. Animals began to die, and hundreds of people fell ill. The smoke coated people's throats and stung their eyes. They could smell it outside in the streets and inside their houses from the time they woke up in the morning until the time they fell asleep.

The Kuwaiti government wanted the wells extinguished as soon as possible. After all, oil was Kuwait's primary industry, and 6 million barrels a day was burning. The country was losing $1,000 U.S. each second. Not only that, but 49,000 tonnes (54,000 tons) of sulfur dioxide were pouring into the air each day. Sulfur dioxide is a chemical that pollutes rainwater, creating toxic showers that pollute rivers, lakes, and fields.

The Kuwaiti government called in four famous North American firms—Boots & Coots, Wild Well Control, and the Red Adair Company, all from Houston, Texas, and Safety Boss, from Calgary, Alberta—to battle the blazes. When the companies' experts arrived in March, they flew over a roiling sea of black smoke, punctuated by burning oil wells throwing flames and fireballs hundreds of feet into the air. More than one firefighter called the trip a journey into hell.

Kuwait hoped to have the wells under control in nine to twelve months. But when the three Houston-based companies arrived, ready to work, they saw that such a task would be impossible. They soon believed that battling the fires could take years. The land between Kuwait City and the oil fields was covered in massive shallow lakes of oil, many of them burning already and all of them ready to explode into flames at any second. As the oil spread, it ignited storage tanks and pipelines until the entire desert seemed pockmarked by fires.

They government needed to build roads across the oil lakes, allowing the teams to approach the burning wells. Kuwait bought hundreds of tractors and cranes and hired teams of welders to work shifts around the clock. Each piece of equipment had to be shielded with corrugated tin to protect it from the intense heat.

Now fire-fighting teams could find their way across the lakes of oil, but the desert itself held danger. Hundreds of Iraqi land mines and Allied bombs lay unexploded in the sand, camouflaged by a coating of oil and soot. And when the crews negotiated safe paths to the fire sites, they found that the emergency and safety equipment stored at many of the wells had been sabotaged or stolen by the retreating Iraqis. There were no tools to fight the fire.

Through March, the firefighters did little except survey the wells and create lists of needed supplies. The government struggled to meet their needs. Even the population of Kuwait City was drinking water from converted tanker trucks, containing remnants of the oil that the trucks usually carried. How were officials going to transport the necessary food and drinking water to the crews in the desert? Each airstrip would have to be searched for mines and bombs before use.

Even beyond transportation, minesweeping, and equipment needs, there was one overwhelming problem—water. The oil fields lay in the center of a desert, and dousing a 4,000-degree flame required millions of gallons of water. How could they get the necessary water onto the sand dunes?

The most common method of dampening an oil well fire was to arrange plastic explosives around the well. Detonating the explosives would cut off the oxygen supply to the oil fire for a fraction of a second. That fraction was all firefighters needed to begin spraying the well with water. Eventually, the streams of water would cool the metal components of the well and the surrounding sand and prevent the well from reigniting. Only then could the crews move in and try to control the gush of pressurized oil erupting from the earth. For each well, a small lake's worth of water was needed.

The government hired Bechtel Corporation to somehow supply the firefighters with the necessary water supplies. Bechtel began digging massive lagoons at the beginning of April. By the end of the first week, they had completed three plastic-lined storage tanks capable of holding up to 4.5 million liters (1.2 million gallons) each. Then they converted a pipeline to run backwards. Instead of carrying oil from the desert to the Persian Gulf, it would now carry seawater from the Gulf to the new lagoons. It could carry 18,000 liters (4,700 gallons) of seawater each minute—enough to fill an Olympic-sized swimming pool in less than four hours.

The firefighters didn't wait for the pipeline to be completed. The Safety Boss crews arrived on April 7 and began fighting some of the smaller well fires with methods that didn't require such large amounts of water. The company president, Mike Miller, had decided that being able to move in and out of well sites easily would be the key to their success. He brought his own 227,000-liter (60,000-gallon) water tankers with him. Instead of plastic explosives followed by bursts of water, his crews poured chemicals onto the wells to put out the flames. It was more dangerous and only worked on the smaller fires, but the method needed much less water.

Boots & Coots began a similar process with liquid nitrogen. They attached a hose to a vertical steel tube. Using a tin-shielded crane, the crew then lowered the tube over the well, spraying it constantly with water to prevent it from melting in the heat. When it was positioned perfectly, the crew started an engine on the ground that heated the liquid nitrogen, turning it into gas. The gas traveled through the hose, down the vertical tube, and over the top of the well. It displaced the oxygen needed by the fire and doused the flames. Water was still needed to cool the machinery, but not nearly as much. After the flames were out, the crew poured a thick mixture of mud onto the well, forcing it closed.

In mid-April, when Bechtel Corporation succeeded in supplying seawater, crews were able to switch to the faster method of plastic explosives. By April 12, 32 wells had been capped. By April 26, with the improved water supplies, crews had capped 60 wells.

Still, the oil fields were far from safe. The firefighters were constantly alert, aware that an explosion could happen at any second. On April 25, a journalist and a photographer were killed when their car slid into an oil lake at the side of the road. The heat from the car's engine ignited the oil and the men burned to death. Three other men were killed in similar incidents on the same day.

Despite these accidents, the crews continued to battle. Together, the four companies capped 52 wells in May, 48 in June, and 68 in July. Hoping to extinguish the fires even more quickly, the Kuwaiti government called in fire-fighting crews from around the world in September and October. A Chinese crew capped 10 wells, a Kuwaiti team capped 41, and a French crew capped 9. A Hungarian crew arrived with an invention that sparked some mockery and some admiration among the firefighters. The team had outfitted Russian tanks with jet engines, allowing them to blast water and fire retardant toward the fires. They managed to cap nine wells.

At the height of the battle in September, kitchens were providing up to 26,000 meals per day for firefighters. The Kuwaiti government had accumulated so many cranes, bulldozers, backhoes, tankers, and other pieces of heavy equipment that the country owned the world's largest fleet of non-military vehicles.

Finally, on November 8, Safety Boss capped the last burning oil well. It was their 180th well. The Red Adair Company had capped 111, Boots & Coots had capped 128, and Wild Well Control had capped 137. Each team had completely extinguished all fires in its assigned sector in under nine months. It was a feat that astonished scientists and experts from around the world. The effort had cost the Kuwaiti government $1.5 billion, but it had also saved the future of the nation's primary industry.

Smokescreen

The fires were out, but the environmental devastation remained. The smoke from the oil well fires traveled around the world. It was found in snow over the Himalayan Mountains between India and Nepal. It was detected in the atmosphere over Japan. Scientists at the observatory on top of the Mauna Loa volcano in Hawaii found oil smoke in their air samples. Some meteorologists believe that these particles changed the earth's air temperature in some places and contributed to dramatic weather patterns in 1991, including a cyclone in Bangladesh, flooding in China, and blizzards in Israel.

Plants and animals also felt the effects of the Gulf War. Usually the mangrove thickets and salt marshes of the Kuwaiti coastline supported fish, shellfish, invertebrates, and waterfowl. When Greenpeace surveyed the area in August 1991, they found few signs of life. Only a couple of mussel and worm species, able to live in highly polluted waters, were still alive. In the years after the last well was capped, local governments spent more than $700 million on environmental cleanup.

Environmentalists still monitor the region, but the world's attention quickly moved on after the last oil well was capped. Journalists flew out of the country, chasing new stories, and the fire-fighting crews dispersed. Kuwait was left on its own to clean up more than 200 lakes of spilled oil, mop up a petroleum-coated coastline, and dredge almost 1,000 square kilometers (390 square miles) of desert—the sand was buried under a layer of soot and ash. Iraq had failed to destroy Kuwait's petroleum industry, but the fires had caused one of the worst environmental catastrophes of the twentieth century.

London Is Burning

London, 1666

BANG, BANG, BANG! A hammering woke Thomas Farriner in the early hours of the morning on September 2, 1666. What was it? Was the wind swinging the iron bakery sign outside? Was someone at the door of the shop downstairs? *Bang, bang, bang!* There it was again, even louder this time.

Farriner sat up in bed but soon fell back on the mattress, coughing. The room was filled with smoke ... something was on fire! Dropping to the floor where the air was clearer, he scrambled toward the hallway. There he found his panicked servant pounding at the door. Together, the two men rushed to wake Farriner's daughter and the maid.

They had to escape. Farriner led the frightened group to the stairs, but flames were already licking the handrails, and a blast of heat rose to meet them. Then he tried a room at the front of the house, but flames poured out, singeing Farriner's daughter and sending the maid into hysterics. Finally he decided they would have to crawl out onto the roof.

Farriner was the king's baker and like many merchants of the time, he lived above his shop. He was responsible for supplying the royal navy with ship's biscuit, a dry flatbread made of wheat. The English were at war with the Dutch in 1666 and Farriner's ovens were constantly burning. He and his staff were responsible for supplying each sailor in the navy with a pound of biscuit per day.

Farriner struggled to remember the evening before. He was positive that he had personally checked each oven, to be sure the flames were extinguished. How had this fire started?

The houses and shops along Pudding Lane, where Farriner lived, were built several stories high. Many were old, and they leaned precariously over the narrow street below. Farriner carefully stepped out onto the slippery roof tiles, a strong wind pulling at his nightclothes. Once he was balanced, he reached in to help his daughter and the servants. The maid didn't move, so Farriner tried to reason with her. She would be burned to death if she stayed, he called. Huddled against the wall inside, her eyes wide with fear, she still refused to join him.

The flames crackled closer, orange tongues reached from other windows onto the roof, and eventually Farriner had no choice—he was forced to leave the maid behind. He led his small group carefully over the eaves and onto the roof of the neighboring house. There they pounded on a window until the family awoke.

Soon the street was in chaos. "Fire! Fire!" men yelled, banging on drums to spread the warning. A few people ran to the local church for the fire-fighting equipment stored there. They found a few leather buckets, ladders, and the fire hooks that were sometimes used to pull down burning buildings. Much of the equipment was in disrepair.

While families hurried to safety and some men struggled to move their valuables, Farriner and his neighbors formed a bucket chain. One line of men poured water on the fire while another line passed the empty buckets away. Fire hydrants didn't exist in 1666, but there were pipelines that brought water from the nearby river. There were also a few wells in the neighborhood. Still, fighting a fire by bucket was achingly slow. Despite the men's efforts, the neighbor's house was soon in flames. Farriner's maid was killed, the first to die in the fire.

A Seventeenth-Century City

Though the bubonic plague had ravaged the population the year before, London in 1666 was still a bustling city. It was one of the largest business centers in Europe, a major port, and home to a flourishing culture. But while politicians tried to devise better ways of governing the nation and wealthy merchants worked to conquer new areas of the world, hundreds of thousands of poor people lived in crowded alleys, where sewage flowed through open channels and women emptied garbage pails or chamber pots right onto the streets. Often several families lived in one house, which allowed diseases such as the plague to spread quickly. These houses were packed almost on top of one another, lining narrow lanes. A fire could spread in an instant.

The man who was supposed to govern both the rich businessmen and the poor peasants of London was Sir Thomas Bludworth, the Lord Mayor. Many people believed that Bludworth was uneducated and indecisive and that he was governing London badly. But he had supported King Charles II at a time of upheaval, and the king had rewarded him with control of the city.

Bludworth was unhappy when his advisors woke him at 3 a.m. on Sunday, September 2, to tell him about a fire at the home of the king's baker. He was irritated when they advised him to call for a carriage and survey the fire. And he was furious when his carriage couldn't get through narrow Pudding Lane. He was forced to step down into the muck of running water and debris and wade through the crowd of onlookers to get close to the fire.

"Bah," he spat when he finally saw it. "A woman could piss it out." With that, he returned home and went back to bed.

Sparks Fly

After Bludworth left, the fire spread across the street to the Star Inn and its stables. Soon the inn's guests, the staff, and the livestock were fleeing down the street. The fire fed on the stable's straw and dry timbers, then raced down the lane, consuming flimsy shops and houses as if they were matchsticks. In the cellar of a nearby shipbuilder's shop, barrels of pitch—used to waterproof the hulls of boats—exploded. The winds quickly picked up the debris and carried it across the neighborhood. The local church, St. Margaret's, was soon in flames.

There was pandemonium in the streets. Drumbeats and ringing church bells continued to warn citizens. Families scurried back and forth from their homes, trying to move their belongings through the dark, smoke-filled streets. There was no fire insurance in 1666, so anything lost would be gone forever. Many people stood in their doorways until the last possible second, hoping the fire would spare their homes. They were forced to run from the flames, hearing the crack of timbers behind them as their houses burned.

Among the markets and shops of the area, someone found a wheeled water pump—one of a few dozen stored around the city at that time. If a chain of people with buckets filled a tank in the pump, workers could use it to send water through a hose and onto the flames. However, the pump couldn't spurt water very far and soon the fire was too hot for the machine to get close.

Again, the mayor's advisors woke him and told him about the fire's progress. This time, as the mayor surveyed the spreading flames, his advisors begged him to take action. He must order a row of houses pulled down, they said, and have the debris cleared away. This would create a wide open space—a firebreak that the flames couldn't cross.

The mayor refused. Who would pay for the destroyed homes? According to law, they would be his responsibility. He didn't want to be sued by the homeowners of London.

The flames crept closer and closer to the docks along the River Thames, devouring timber-framed shops and warehouses along the way. As the fire reached each new building, it found more to feed on—barrels of brandy or tubs of tar, wharves stacked with timber and coal. By breakfast time on Sunday, London Bridge itself was on fire.

Diary of Destruction

A hillside away, Samuel Pepys ate a leisurely breakfast. As a clerk in the navy, Pepys was level-headed and efficient, dedicated to his work. But he had a second, secret interest—his diary. From 1660, when he was 26, until 1669, Pepys recorded most of his experiences in detailed journals. His observations have allowed modern historians to understand what life was like in London in the seventeenth century. One of his most exciting volumes contains his descriptions of the Great Fire of London in 1666.

Pepys had seen the fire from an upstairs window about 3 a.m. He was interested, but the blaze seemed small and far away. Then, after breakfast on Sunday morning, his maid told him the flames were spreading. Three hundred houses had been burned.

Alarmed, Pepys walked to the Tower of London, which offered a panoramic view of the city. The flames were definitely spreading. He could see them fanning out in a wide arc and he could glimpse the charred and blackened area behind them. Still curious, Pepys hired a boat to take him onto the River Thames. From the water, he watched as people threw their possessions in the river to save them or piled their belongings onto boats or rafts. Pepys watched for an hour and saw no one attempting to stop the flames. Families concentrated on saving their belongings and escaping the area.

When he had seen enough, Pepys traveled to Whitehall Palace, home of King Charles II. There he found some of the king's courtiers in the chapel. When he told them of the fire, the courtiers arranged for Pepys to meet King Charles II and his brother, the Duke of York.

Charles II was known for his gambling, drinking, and womanizing. He used the luxuries of the palace to better entertain his mistress and his numerous girlfriends. Even though he was more of a playboy than a king, he recognized the danger of the fire. If it wasn't stopped, it could reach Whitehall Palace, the base of his power. Charles II listened carefully to Pepys's news, then sent the clerk with a message to the mayor: pull down any homes necessary to create a firebreak, and let the homeowners sue the king. The king's brother, the Duke of York, added that the mayor could request any soldiers needed for the task.

Bursting with importance, Pepys set off to find the mayor. He wove through streets crowded with fleeing families and carts piled with possessions. Some men bore huge sacks on their backs. Others carried stretchers, shepherding ill or elderly people to safety. Many people who had escaped their burning homes a few hours before had moved their families' belongings to the homes of friends and relatives. Now, more and more neighborhoods were ablaze. Forced to move again, peasants and workers chose their directions haphazardly. The river was already choked with debris. The fire attacked the shanties of the poor and the homes of the wealthy with equal vigor. Nowhere seemed safe.

Weaving through the chaos, Pepys made his way closer to the front lines of the fire. Houses crackled as if they were made of kindling, tall buildings turned into blazing torches, and smoke billowed into the sky. Pepys found the mayor nearby, mopping his brow. He had finally tried to create firebreaks. The men had started too near the flames, however, and before the houses could be completely demolished, the fire had arrived. The intense heat had driven the workers back.

When Pepys repeated the king's orders, the mayor looked like he was about to faint. He was exhausted, he declared, and pulling down houses was never going to work. He didn't need soldiers—he needed rest. He hadn't slept well the night before, and he was going home to bed. The mayor rode away. Behind him, the flames continued to rise. Men struggled to pull carts loaded with all their household goods. Children wailed as they were whisked from their homes.

Who's In Charge?

On Sunday afternoon, King Charles II and the Duke of York sailed down the river to see the fire for themselves. The king immediately gave orders to have firebreaks constructed, but the flames seemed impossible to stop. As the wind fanned them, they ignited one building, then another, a boiling force reaching steadily up the hillside toward the center of the city. By Sunday night, the fire was visible from the palace windows.

By the morning of Monday, September 3, it was obvious that the mayor wasn't going to take charge. The king gave his brother, the Duke of York, control of the city. Quickly, the duke picked his most level-headed noblemen and placed each in charge of a fire station. The stations were arranged in wide arcs around the blaze.

Average London residents had little motivation to fight the fire. If their houses and workplaces were already destroyed, then it seemed more important to escort their families to safety. Trying to convince people to stay and help, the duke promised rewards from the king. The duke also called in sailors from nearby ports to help pull down homes and create new firebreaks.

For the next three days, workers valiantly battled the blaze. Despite their efforts, the fire spread. It roared down Gracechurch and Lombard streets. It tore through the Royal Exchange, where it destroyed stacks of the finest silks and luxury cloth and the East India Company's stores of expensive, imported pepper. Homes that had once sheltered up to 10 poor families went up in minutes, and the posh mansions that lined the streets of the Cheapside area burned just as quickly. The Custom House burned, and the Tower of London was threatened. So much smoke poured into the air that it could be seen from the Scottish border.

As the fire destroyed more districts, soldiers, sailors, and dockworkers began a huge demolition project. They placed kegs of gunpowder in a row of buildings, then connected the kegs with lines of explosives. The entire row blew up in a series of earth-shaking blasts. The men had barely finished the firebreak, however, when strong gusts of wind blew the flames right over it to the other side.

Some people compared the flames to the fires of hell. Shacks and shops, factories and infirmaries disintegrated in seconds. Many people had thought the massive stone walls of St. Paul's Cathedral would protect it, and they had placed belongings inside for safety. Unfortunately, the piles of clothes and keepsakes inside caught fire and the air inside the church grew hotter and hotter. The temperature rose so high that the massive stones in the walls began to burst like grenades.

When he wasn't directing the firefighters, the duke led a group of guards around the city, controlling crowds and keeping people calm. Both the duke and the king appeared regularly at the fire stations, encouraging workers and handing out coins as rewards. The king himself sometimes joined the bucket chains and was soon as blackened and dirty as the workers. The duke worked even harder, staying late into the night. At one point the wind carried sparks directly over a firebreak in progress, and the duke found himself surrounded by flames. He and his men were forced to run through the heat and smoke to save themselves.

Just when it seemed the fire would never be over, there was another crisis. No more water! In the previous days, people had tapped into the pipes under the city, needing water to try to fight the blaze. Their fire-fighting efforts had been unsuccessful, but so much water had drained away that the city's water system was now dry and useless. Building firebreaks was the only strategy left and by now, even the soldiers were exhausted.

Changing the Tide

Finally, on Wednesday morning, the winds calmed. The flames moved more slowly and firefighters began to hope. Residents who had fled the city began to return to help the efforts. Even the mayor appeared again, directing the creation of a firebreak.

Late Wednesday night and in the early hours of Thursday morning, the flames were beaten. Without the wind, the fire crews were able to contain the blaze to the areas already burned. There, the fire soon ran out of fuel. Only charred heaps and smoking remains of houses were left behind.

The flames had subsided, but many devastated, homeless people roamed the ruins. Seeing that there were few stocks of food left in the city, the king ordered the surrounding counties to send any spare supplies to London. He created temporary markets around the fire site and ordered all remaining churches, schools, and public buildings to serve as storehouses. He even opened the navy's storehouses and made the reserve stocks of ship's biscuit available to the public.

Rumors swept the city that the fire had been the work of England's enemies—the Dutch or the French. Men who had spent days hungry, sleepless, and terrified suddenly turned violent. Mobs began hunting for scapegoats and beating any immigrants they found. Crowds pressed in around the palace. Only a speech by the king himself finally calmed the rioters. "The judgment that has fallen upon London is immediately from the hand of God," the king told them. They couldn't blame immigrants or England's enemies—they had to blame themselves.

As the people gazed over the flattened city, they believed that God had punished them. Five-sixths of the city—more than 80 percent—was gone. More than 13,000 houses had burned and 100,000 people were left homeless. Eighty churches were gone and some people couldn't even find the lots where their homes had been— the landmarks had disappeared and the neighborhoods were covered in smoking soot and ash. One observer was shocked to see that the bells of St. Paul's Cathedral had melted in the heat.

Recovering

Amazingly, only six people had died in the blaze, and London's wealthy residents managed to rebuild quickly. Samuel Pepys was thrilled to find his home, his office, and his diary pages intact. Sir Thomas Bludworth, the mayor, lost his house, his reputation, and eventually his position. But he kept his wealth and replaced his burned home with an even more splendid mansion.

Most Londoners weren't so lucky. It took the city's workers and business owners years to rebuild, and some were forced to move to other towns or cities to find work. Many of the poor had lost everything and found themselves even more destitute than before.

Thomas Farriner, the king's baker, testified as part of an investigation after the fire. Determined not to be blamed for the destruction of the city, he insisted that all of his ovens were cold on the night of September 1, 1666. He had even tried to light a candle from one, he said, and there hadn't been enough coals. Although many believed that the baker's ovens must have started the blaze, this was never proven.

In a final act of leadership, King Charles II issued a royal decree after the fire: all buildings should be rebuilt using stone or brick, protecting London from future fires. Those who remembered the days and nights when houses blazed, churches were transformed into furnaces, and warehouses and docks collapsed into ruins were anxious never to allow such a disaster again.

Death Train

Taegu, South Korea, 2003

KIM DAE HAN STOOD JUST INSIDE THE DOORS of the subway car, a black bag slung across one shoulder and a lighter in his hand. It was 10 a.m. and the morning rush was over, but the train car was still crowded. College and university students in brand-name gear swayed slightly to the music from their earphones or chatted on their cell phones. A few conservatively dressed businesspeople stood silently near the doors. Older women carried shopping bags on their arms, heading downtown to tour the stores.

Fifty-six-year-old Kim Dae Han began to flick his lighter on and off, on and off. He hated these people. He hated everyone. He glanced around the train car in disgust. He didn't used to have to take the subway. Prior to his stroke two years before, Kim Dae Han had driven his taxi through the maze of streets above. He knew his way around every neighborhood, from the blocks of textile factories to the fashionable streets lined with boutiques. He had often spent his days shuttling businesspeople to and fro between the buildings of the Financial District, where this subway train was headed.

Kim Dae Han suffered from mental illness because of his stroke. On some days, he thought it was the hospital's fault for not taking proper care of him. Once he had even gone to the hospital and threatened to burn it down, screaming at the staff until the police had arrived. He was still angry with the doctors and the nurses, but now he had added more people to his list. Those who negotiated the city so easily, when it was so difficult for him. The ones who bumped or jostled him in the street. The ones who were impatient when he took too long to count his change or collect his things.

That morning, February 18, 2003, he had decided there was really nothing useful he could do with his life anymore, and the best solution he could think of was suicide. But he didn't want to die alone. In his mixed-up view of things, other people should die with him. So he stood at the edge of the subway car, flicking his lighter on and off, waiting for a moment of courage. He drew a milk carton from his black bag and held it close to his chest.

Suddenly an elderly woman shoved closer to him. "Stop that!" she said. Kim Dae Han ignored her, focusing only on the lighter flame.

"Stop that!" she said more loudly. Then she pushed him.

That was all the encouragement Kim Dae Han needed. As the train pulled into Jungangno Station, he held up his milk carton—filled with gasoline—and dropped in the lighter.

Life in Taegu

South Korea's third-largest city, Taegu is home to 2.5 million people. It is surrounded by valleys and is famous for its apples and medicinal herbs. The city is also the base of Korea's textile industry. In local mills and factories, workers spin thread, weave cloth, and produce clothes from sportswear to luxury furs. Some of the products made in Taegu are sold in the prestigious fashion boutiques of Milan, Italy.

Though less cosmopolitan than South Korea's capital city, Seoul, Taegu has a bustling downtown full of modern buildings.

Several decades ago, the city government began widening streets, creating parks, and encouraging innovative architecture, things that have helped make Taegu look like a modern metropolis.

The subway line was another part of this drive to modernize—the 30-kilometer (18-mile) system was completed in 1997. The local people had immediately taken to the trains, eager to escape the city traffic and travel in the sleek, high-speed railcars. Unfortunately, the rush to improve the country had sometimes led to low safety standards. An explosion during the construction of the subway in 1995 killed 101 people and led some journalists to call South Korea "the republic of accidents."

Panic Strikes

When Kim Dae Han dropped his lighter into his container of gasoline, flames instantly erupted from the top of the carton, searing his hand and climbing his sleeve. Men around him noticed the fire and leapt on top of Kim Dae Han, crushing him to the ground in an attempt to smother the flames. It didn't work. By this time, the train had come to a complete stop and the doors had slid open. Soon the flammable tiles of the subway car's floor were ablaze and smoke spread through the car and onto the platform, choking the passengers as they tried to escape. Kim Dae Han felt fumes scorch his lungs as he struggled out from under the men and scrambled toward the door. His clothing blazed as he raced for the exit.

People who had been standing on the platform, waiting to board the train, panicked and fled. As the remaining passengers also ran toward the exits, the car they had left erupted into a fireball, quickly expanding into a globe of raging red. Tongues of flame reached out from the doors as if searching for new fuel. The vinyl on the train seats caught fire, sending toxic fumes pouring into the air and growing so hot that the aluminum framing of the chairs and handrails soon began to buckle and twist. The plastic of the safety straps melted like wax, dripping to the already flaming floor below.

Passengers in the other five cars of the subway train didn't see what had started the blaze, but many saw the flames erupt and smoke burst from the doors. They ran for the exits without waiting to investigate. Behind them, the air within the car grew hotter and hotter until the fire was able to burst through the frame and attack the cars on either end, turning them almost instantly into raging infernos. Soon the blaze burned through the metal of those cars as well, and the fire began to leap from car to car until all six were alight. By that time some of the passengers had made it to the stairs, scrambling up toward the street. Others were still trapped below, trampled by the crowd or overwhelmed by the toxic fumes of burning plastic before they could find the exits.

The Next Station Is . . .

Mr. Choi had driven subway trains for several years without incident, but as his train pulled toward the Financial District, his radio crackled to life.

"When you enter the Jungangno Station, drive carefully. There is a fire," said a subway controller.

Obediently, Mr. Choi slowed his train as he approached the station. Then he saw the scene ahead of him—black smoke billowing over the platform and the twisted, half-blackened skeleton of the train on the opposite tracks, still glowing with flames. A wave of heat seemed to wash over him. The fire was obviously much larger than the controller had expected. Mr. Choi stopped his train at the platform but kept the doors closed as he radioed for more instructions.

"It's a mess. It's stifling. Take some measures, please. What should I do?"

The controller was unable to direct him, and Mr. Choi soon found that his train was trapped. Power to the rail lines had been cut, and he could no longer reverse down the track. "Stay calm," he told the passengers over the intercom system, leaving the doors closed to prevent the smoke from entering the cars.

Chang Ki Ja was on board Mr. Choi's train with her five-year-old son. As the train had slowed and then begun to creep toward the platform, she and the other passengers had realized that something must be wrong. They soon saw the thickening black smoke and the raging fire itself and heard Mr. Choi's voice urging them to remain calm. For a few moments, people stood quietly. Then the fumes began seeping in around the edges of the closed doors and windows. The air seemed to be growing hotter. Somehow passengers managed to force open the doors of the train car. Clutching her son, Chang Ki Ja tumbled onto the platform and stumbled through the smoke to find an exit.

Behind her, the passengers of five other cars remained trapped, coughing and wheezing as the smoke roiled around the train. They tried to break the seals on the safety doors, but no one knew that the emergency lock releases were hidden underneath the seats. There were no safety instructions on the doors. Some people screamed and pounded on the windows while others huddled in shock against the hot metal walls.

Suddenly the lights went out. Fire doors around the platform slammed closed. From the control room, engineers had cut the power and implemented emergency fire procedures, hoping to keep the fire from spreading to the other three levels of the station. But they left hundreds of passengers trapped in the dark. Emergency lights on the station's ceiling flickered to life but were blocked by the black smoke. The ventilation system had also been connected to the main power supply—when the lights went out, the fans went off. The smoke grew even thicker and flames began to break from the first train and blaze across the platform.

Mr. Choi, the driver of the second train, remained inside his car for 20 minutes, until the fire was almost upon him and it grew difficult to breathe. He could hear the panicked screams of people in the cars behind him and he could feel them pounding against the train walls. He radioed again to the control room. "What should I do?"

"Switch off the car and run away," the controller said.

Mr. Choi did just that, frantically sprinting across the platform to the emergency stairs, leaving the doors of the train cars still locked behind him and the passengers trapped inside.

Calls for Help

Chung Hyun Suk worked as a dispatcher at the Taegu Fire Department. Just after 10 a.m., she answered an emergency call from Jungangno Station. "People are trapped in the subway," a woman's voice said. "They are suffocating to death. Please come quickly and open the door!"

The call cut off, but the phone rang again almost immediately. It was a different woman. "A fire broke out in the subway. Please come right now. I'm going to die!" she cried.

Chung Hyun Suk sounded the alarm and sent details to the firefighters on call. She continued answering panicked pleas for help, one after another, each caller sounding more desperate than the last. Then, 20 minutes after the first phone call, the lines fell silent. She waited, wondering. Had firefighters arrived? Or had the last trapped person succumbed to the smoke and flames?

Chung Hyun Suk wasn't the only one to receive calls from

the scene. Many of the subway passengers were carrying cell phones, and they spent the last few moments of their lives saying good-bye to their families.

"Mom, a fire broke out," said high-school student Park Nam Hee after managing to speed-dial her phone in the dark.

"Don't be overwhelmed," her mom told her. "Stay calm." Shortly after, the phone line went dead.

Mrs. Lee received a call from her daughter Lee Sung Young, a 20-year-old student: "Mom, there is a fire in the tunnel. I think I'm going to suffocate and die here. Mom, I love you."

Where Were the Rescuers?

Soon after dispatcher Chung Hyun Suk received the first emergency calls, fire trucks and ambulances were speeding toward the station, their lights flashing and sirens blaring. As they neared the area, they could see smoke pouring onto the street from subway ventilation ducts. Drivers had stopped to investigate, a few train passengers had already escaped to street level, and a crowd was quickly gathering. As the rescue vehicles joined the fray, the road was almost completely blocked, creating a massive traffic jam in the downtown core.

While ambulance workers tended to the passengers above ground, firefighters plunged into the station. The upper levels were relatively clear, but the smoke in the stairways was impassable. The rescuers were soon retreating to street level, choking on fumes. They called for shields and oxygen tanks and waited helplessly outside. Eventually they managed to enter again and carry some of the passengers trapped in stairwells and hallways to safety. The burning platform itself was still too dangerous to approach.

The Aftermath

At 1 p.m., three hours after the blaze had erupted, the fire slowly ran out of fuel and oxygen. Rescuers wearing gas masks and protective clothing were finally able to access the station platform. It looked like the blackened shaft of a coal mine. Soot and smoke mixed in the air to form a haze over wall advertisements and subway signs, things that now looked strange above the wreckage. The twisted train cars were almost unrecognizable—only their metal frames still slumped haphazardly along the track. On the platform itself, bones lay scattered among scorched purses and shoes. Rescuers found many bodies in front of locked fire doors or trampled in narrow stairway entrances.

Within days, soldiers were enlisted to shovel the debris from the platform and tracks. As they worked, a group of relatives set up camp in the charred rubble nearby, waiting and hoping for the remains of their missing relatives to be pulled from the wreckage. Park Kyung Woo dragged a mattress and blankets to the site and resolved to wait until the body of his mother was found. When he had last spoken to her, she was about to take the subway to church. Now he was sure she lay buried in the station. And he told reporters that her death was especially painful, because he had worked on the crews that built Jungangno Station in 1995. "I feel I dug a grave for my mother," he said.

As the days passed, the death tolls reported by the newspapers rose from 20 to 125, 130, 133, and finally to 196. Many of the bodies were badly burned, and investigators collected DNA from family members to try to identify the victims. More than a hundred other train passengers lay at home or in the hospital, recovering from burns or respiratory injuries from the toxic gases.

As public outrage grew over the accident, the driver of the second train, Mr. Choi, turned himself in to police and was charged with negligent manslaughter. The head of the subway system was fired, and four subway controllers and the driver of the first train were arrested and charged with negligence. Kim Dae Han, who had started the fire with his lighter, was treated in the hospital for severe burns. He was arrested there by police and was later sentenced to life in prison.

He had come to hate humanity, he told police officers. "I want to die, but I want to die with a lot of people," he said.

In the months after the disaster, investigators blamed the extent of the damage on a combination of poor safety measures on board the trains and bad decisions by the controllers and drivers. The South Korean government promptly doubled the budget for subway fire prevention. Meanwhile the operators of subway systems around the world examined their own policies. They also kept watch for possible copycat crimes. They worried that even the highest safety standards might not be enough to counter the deadly mix of one suicidal person and a mob of panicking passengers.

Airship Aflame

Lakehurst, 1937

CAPTAIN ERNST LEHMAN STROLLED PROUDLY along the promenade deck while excited passengers gathered at the windows. Women in long, slim dresses and strings of pearls chatted in small groups, while the men took nonchalant poses, dapper ties tucked neatly into their vests and handkerchiefs folded in their suit pockets. Weaving efficiently through the crowd, crew members offered cocktails or information about the ship. The vessel had just set sail. But this was no luxury cruise liner. This was Germany's finest airship—a blimp.

There were no passenger jets in 1937, but there were plenty of people willing to travel. On tour company walls, illustrated posters advertised the pastoral beauty of Scotland, the architecture of Italy, or the nightlife of New York. For the wealthy of Europe and North America, trips across the Atlantic were a must, and what better way to travel than an airship?

Though Captain Lehman could hear the distant thrum of the engines, he felt no vibrations, no swaying. The lighter-than-air craft seemed to glide above the ground. Only a slight breeze from the open promenade deck windows told him the *Hindenburg* was moving. He couldn't help smiling. He was the director of the Zeppelin Company and one of the most famous airship pilots in the world, and the *Hindenburg* was the company's greatest achievement.

Blimps and Airships

Today, everyone recognizes the Goodyear Blimp as it floats high above sports stadiums, offering bird's-eye views to TV cameras. Shaped like a football, this huge sack filled with helium gas can glide gently through the air. From within the cabin or control room attached to the bottom, crew members steer the blimp and control the engine.

The Goodyear Blimp is a descendant of the airships developed in Germany in the early 1900s. These giant vessels had rigid frames made of lightweight aluminum. A rubberized canvas was stretched across the frame, then the sacks, or cells, inside were filled with helium.

Count Ferdinand Zeppelin was a pioneer in the field. In fact, airships were often called zeppelins in his honor. His firm, the Zeppelin Company, built about 130 airships between 1900 and 1937. When he tested his first airship in 1900, a young news reporter named Hugo Eckener was watching from the ground. Eckener was so excited about the massive flying machine that he immediately quit his job and joined Zeppelin's company, becoming the count's protégé and eventually the company director.

By 1911, airships were carrying passengers across Europe. Then, during World War I, the Germans used airships to carry bombs over Britain. The British suffered huge losses for several years, until they devised ways to shoot down the airships. They also started building their own.

In 1919, a British airship called *R-34* made the first crossing over the Atlantic Ocean, traveling from Scotland to New York City in just over 108 hours. In a time before passenger planes, this was unheard-of speed! Within a few years, airships were offering hundreds of upper-class passengers a faster, more comfortable way to travel. Instead of suffering from seasickness for two to three weeks aboard a ship, passengers could float across the ocean in less than three days.

Fabulous Fish

Hugo Eckener was transfixed by airships. When he took over leadership of the Zeppelin Company, he was convinced that the vessels could do more than merely cross the Atlantic Ocean. In 1929, he set out on the most ambitious airship journey of all. In the *Graf Zeppelin*, he traveled around the world. The 21-day voyage took him to Tokyo, San Francisco, and Lakehurst, New Jersey. He returned to Germany triumphant, calling his airship "a fabulous silver fish, floating quietly in the ocean of air, and captivating the eye."

Eckener loved the way the airship made crowds stop and point to the sky. He considered the vessel a symbol of peace and a way of uniting far-flung regions of the world.

When Eckener returned from his record-breaking voyage, he set out to build the biggest, most luxurious airship of all. And he did it. The *Hindenburg* was longer than two football fields—almost as long as the *Titanic*. It was as tall as a 13-story building. The airship boasted a high-tech control room, comfortable passenger cabins, fine-dining facilities, a reading lounge, a bar, and a promenade deck from which passengers could enjoy the view.

The only problem was the gas needed to fly the *Hindenburg*. The ship was originally designed to use helium. But Adolf Hitler had taken power in Germany, and the United States refused to sell helium to German companies. Eckener was forced to use a more flammable gas—hydrogen—to lift his airship. He took every precaution necessary to avoid fire. He coated stairways and catwalks in rubber and ordered the crews to wear soft-soled shoes, so no sparks of static electricity could ignite the gas. He also had passengers' matches and lighters confiscated before takeoff. People were allowed to smoke, but only in a pressurized room with an air-lock door, where the steward controlled the only lighter on board the ship.

In 1936, the *Hindenburg* completed six round-trips between Germany and Brazil and 10 round-trips between Germany and Lakehurst, New Jersey. In total, it carried 2,656 passengers. A one-way flight cost $400 U.S.; a return fare was $720.

Despite the fanfare and luxury of these journeys, the Zeppelin Company was facing political problems. Eckener, as the company chair, was a vocal opponent to Hitler. When Hitler demanded that the airships carry the swastika, the sign of his political party, Eckener was unhappy. And when Hitler had the *Hindenburg* used for a propaganda event, Eckener was furious. Because of his protests, he was asked to step down as the company director. An experienced airship pilot, Captain Ernst Lehman, was appointed in his place.

Across the Atlantic

On May 3, 1937, the *Hindenburg* set out on a routine voyage from Frankfurt, Germany, to Lakehurst, New Jersey. There were 36 passengers and 40 crew members on board. There were also 21 Zeppelin Company staff members, many of them training for future airship voyages. Although Captain Max Pruss was in charge of the journey, Captain Lehman had decided to make the voyage as well.

The *Hindenburg* motored across the Atlantic in two and a half days, withstanding ocean storms with only a few shudders. But when Captain Pruss approached Lakehurst at about 4 p.m. on May 6, he was told that thunder and lightning would delay his landing. Because of the flammable nature of the hydrogen in his ship, he couldn't land when there was any danger of lightning.

Captain Pruss waited until 7 p.m., when the dark clouds had started to lift, before he approached the landing field. He made a sharp turn, lowered the ship, and dropped the landing ropes. The ship was 23 meters (75 feet) in the air.

Suddenly, just as members of the ground crew were rushing to grab the landing ropes, a streak of flame burst from the top of the airship. Within seconds, a huge mass of fire and smoke was rising from the tail. First the flames looked like an enormous balloon, then they rose into a roaring mushroom. The once luminous ship became a bomb, sending fireworks in every direction as explosions rocked the vessel and the stern came plunging toward the ground.

Radio reporter Herbert Morrison was on the scene, recording the *Hindenburg*'s arrival for a Chicago radio station. He began by describing the landing airship as light and strong, "a marvelous sight" descending from the sky. As brief glimpses of sun broke through the storm clouds, the airship's windows shone "like jewels on a background of black velvet."

All at once, the recording crackled as the first explosion rocked the earth, and Morrison's voice broke as he shouted, "It burst into flames . . . It burst into flames and it's falling!"

Frantically, Morrison scrambled to describe what was happening. The ship was disintegrating and black smoke billowed outward as flames leapt several stories into the air. "Oh, the humanity," he cried, already imagining the panic within the ship as the wreckage crashed to the ground.

As Morrison continued to record events, the ground crew scrambled for safety. One man tripped as he tried to escape. The burning hull of the airship crashed down on top of him, killing him instantly. As the gas cells in the stern exploded one after another, the bow suddenly shot up, hundreds of feet into the air.

The passengers were sent sliding down the promenade deck or crashing against the walls of their rooms.

Twelve crew members inside the bow clung to support bars as the floor dropped out from under them. A tower of flame shot up from below. Slowly, their hands began to slip. One by one, they lost their grip on the bars and fell hundreds of feet down into the flames. When the bow finally crashed back to earth, only three crew members were still clinging to safety.

Once they learned of the fire, passengers had only a few seconds to escape. Some threw themselves out the windows of the promenade deck. Members of the ground crew risked their own lives to run in close to the flames and escort the passengers to safety.

Trapped near the front of the promenade deck, acrobat Joseph Späh smashed a window and climbed out on the bow. Luckily, he was used to hanging from moving bars! He clung to the metal frame of the *Hindenburg*, high in the air, as the heat of the fire grew more and more torturous. Finally, when he was 12 meters (40 feet) in the air, he dropped to the ground. He broke his heel, but was able to hobble away.

In the control room, Captain Pruss yelled for the crew to evacuate. He knew that a fire on board meant certain disaster. As the ship crashed to earth, seven of the twelve men in the room leapt out the windows. They were all whisked to safety, escaping the blaze as the ship bounced back into the air. When it hit the ground again, the entire vessel already disintegrating from the explosions, the remaining five men jumped. Desperately, they leapt from the inferno of the ship into a disaster scene on the ground, a ruin of twisted metal and burning scraps of fabric, black smoke and searing heat.

The last group to jump included Captain Pruss and Captain

Lehman. Captain Lehman was hit by flaming debris and suffered massive burns on his head and his back. Nonetheless, he was able to walk to an ambulance and talk to rescuers. Captain Pruss was also hit by falling debris but he soon regained his senses. He picked up an unconscious radio officer and carried him to safety, then returned to the flames to search for other survivors. It took three men to restrain him and escort him to an ambulance.

Barely half a minute after the first streak of flame had appeared, the *Hindenburg* lay in ruins. The ground crew continued to enter the wreckage and search for survivors. With their help, a total of 62 people escaped. Thirteen passengers, 22 air crew, and one member of the ground crew were dead.

Captain Lehman was rushed to the hospital for treatment, but his burns were severe. He died the following day. According to some accounts, his last words were "I am going to live." Other people believe he said, "I don't understand it." Captain Lehman, like many other leaders in the airship industry, had regarded the giant airships as miracles of engineering. Unfortunately, this miracle had turned into a fatal inferno.

Wondering Why

What had caused the greatest airship in the world to explode? Reporters, witnesses, and industry experts were all wondering the same thing. In the years after the disaster, three main theories emerged.

The first theory was presented by Hugo Eckener, the former leader of the airship company. Immediately after the fire, American officials called upon Eckener to help them determine the cause. Once he had arrived in New Jersey, heard the witnesses' reports, and surveyed the wreckage, Eckener concluded that the crash was accidental.

He told the media that Captain Pruss had made a sharp turn when he approached the landing field. Eckener guessed that the turn had snapped a support brace. The brace had then ripped open a hydrogen gas cell, allowing gas to enter the airship's top fin. Static electricity, caused by the lightning storms in the area that day, had sparked the gas.

The second theory? Terrorism. Many people, including Captain Pruss and other members of the crew, refused to accept Eckener's findings. Instead, they believed the airship had been intentionally destroyed by a crew member. Because Germany was so proud of the airships and because Hitler had used the *Hindenburg* for propaganda events, igniting it would have been a good way to embarrass the German government.

Among the crew of the *Hindenburg* were several riggers. These men were responsible for walking the ship's ladders and catwalks, inspecting and repairing the cables and the fabric. After the fire, some surviving crew members claimed they had seen a rigger set off a camera flashbulb while standing on a catwalk near the hydrogen gas cells. The brief spark of the bulb would have been enough to cause the explosions. These crew members believed that the rigger was an anti-Nazi activist, determined to sabotage the *Hindenburg*. Captain Pruss was never able to prove or disprove his theory, partly because so many members of his crew—including the suspected rigger—had been killed.

Decades after the disaster, a third theory emerged. American scientists, including a retired NASA engineer, reexamined reports of the accident and discovered that the flames above the *Hindenburg* had been orange. When hydrogen burns, the flames are clear. They also noticed that pieces of the airship's outer fabric remained burning as they fell to the ground.

The scientists were able to obtain two samples of a fabric

similar to the one used on the *Hindenburg*. They analyzed the fabric and found that the coating on it included a highly flammable aluminum by-product. They believed that when static electricity from the surrounding lightning storms reached the ship, the fabric coating could have instantly ignited. It wasn't the hydrogen inside, but the coating outside, that had turned the *Hindenburg* into a massive floating bomb.

End of an Era

With the destruction of the *Hindenburg*, passenger service aboard airships came to an immediate end. Soon the ability of passenger jets to quickly cross the ocean made the blimps obsolete. Eckener, who had dedicated so much of his life to the creation of the vessels, was crushed. He called it "the hopeless end of a great dream."

Yet Eckener also said that he wasn't sorry he had stopped building airships. Hitler had intended to use them as weapons of war, and as the battles began, Eckener was relieved that his ships played no role in World War II. He eventually tried to banish the image of the blazing *Hindenburg* from his mind. Instead, the airships lived on in his memory as grand symbols of peace, the first vessels to fly hundreds of visitors back and forth over the Atlantic Ocean, and pioneers of the new era of air travel.

Operatic Tragedy

Vienna, 1881

IN VIENNA'S RING THEATRE, the audience of 2,000 buzzed with excitement. People hurried up and down the aisles, looking for their seats. Finely dressed Austrian aristocrats were ushered into their private boxes in the balconies above. Workmen scraped scenery pieces across the stage, readying the final touches for the opening act.

It was December 8, 1881, and more than 2,000 tickets had been sold. German composer Jacques Offenbach was already famous for writing more than 80 operettas. Tonight the theatre company would perform his full-length opera, *The Tales of Hoffman*. He had written it just before his death in 1880.

A well-dressed young women had her friend by the hand and was pulling her down the aisle. She loved the theater. As soon as she found their seats, she began surveying the other audience members. She spotted a number of elegant ladies in the balconies and thought she recognized a member of Parliament sitting above her. Around her on the ground level were men who looked like bankers and traders as well as a group of rowdy students causing a commotion at the back. A large family nearby, all laughing and joking, seemed to be celebrating the father's birthday.

It was about 6:45 p.m. and the performance was due to begin in 15 minutes. The young woman reached into her handbag to search for her opera glasses.

Suddenly, from the corner of her eye, she saw the stage curtain blow wildly toward her. She looked up in time to see a wave of flame emerge from under the curtain, instantly enveloping the first rows of seats. For a moment, the audience seemed shocked into silence, then screams echoed through the galleries. Men shouted, "Fire! Fire!"

The young woman and her friend quickly rose to their feet. Luckily they were close to the aisle, and they joined the crowd pressing toward the theater's central entrance. She thought she could hear the crack of the fire behind her and feel its heat on her back. Was it her imagination? Struggling to keep her balance in the pushing crowd, she turned to look.

The theater was ringed by fire as flames climbed the curtains and enveloped the balconies, consuming the ornate velvets and stretching toward the catwalks and rigging above. The stage itself was barely visible, but a great burst of fire, then another, erupted from behind the clouds of smoke.

Just as the young women entered the corridor leading outside, the gaslights lining the theater went dark. The audience was plunged into blackness, and the screaming doubled. The young woman winced and her friend squealed and clutched her arm. Frantically, she pawed at the air around her. She sighed with relief as her fingers touched a handrail. She was at the edge of the corridor. All she had to do was follow the railing to safety.

It wasn't as easy as it should have been. The panicked mob pushed and jostled the pair, almost crushing them against the wall, then picking them up and sweeping them along. Smoke began to fill the passageway. At one point, she realized that her friend was no longer holding her arm.

She called out. A panicked voice answered from nearby, and the two friends were together again in the darkness. In a few more seconds, the crowd poured out onto the street. The young women breathed deeply, their throats scorched and their eyes watering from the smoke. Behind them, they could still hear the screams of those trapped inside.

Behind the Scenes

In 1881, Austrian theatre companies were leading a movement known as Stage Reform. Instead of the simple sets used in the past, the theatres were trying to re-create the look of nature. They used elaborate painted backdrops, equipped with ropes and pulleys. Motors onstage lifted parts of the set up and down as the play required. So that the audience could see and appreciate the complicated set, numerous lamps were usually set about the stage.

With so much flammable material around, lit by gas lamps, fire was an obvious concern. Theaters were required to take several precautions. First, stagehands had to be instructed in the use of an iron curtain. This sheet of metal could be dropped between the stage and the galleries, protecting the audience from any danger. Second, fire wardens were to be placed at intervals throughout the theater, armed with water buckets to extinguish any small blazes that might occur. Finally, the law required that oil lamps, not gas lamps, light the exits. If the supply of gas was turned off, the oil lamps would still show the way to safety.

As the audience arrived on December 8, 1881, the chorus singers had already taken their places behind the curtain at the Ring Theatre. Stagehands scurried in and out with props. A few actors stood in position onstage. The director was there as well, offering last-minute instructions. He was the first one to spot the fire, flaring at the corner of the stage.

Had one of the lamps started the blaze? Had a piece of machinery caused a spark? There was no time to tell. Almost before the director could raise an alarm, the fire was spreading up the ropes and rigging and into the backdrops. A wave of heat struck the actors and stagehands, followed by a billow of smoke. Gasping, they ran for the dressing room at the side of the stage, then out a back door and into the street.

Not one of them stopped to telegraph the fire department. Not one paused to drop the iron curtain between the stage and the audience. And to make matters worse, when they opened the back door to the street, a gust of air rushed in, winding its way through the dressing room and pushing the heavy velvet curtain toward the galleries. The flames followed, escaping the confines of the stage and pouring into the audience area.

There was almost no time to react. As people scrambled over one another to escape, smoke and flames consumed rows of seats, climbing the ornate woodwork to reach the balconies above. Orange tongues shot toward the theater dome and soon the roof was ablaze, sending showers of sparks down on the fleeing crowd.

The audience members on the ground level were the most fortunate. They had a direct escape route through the theater's wide main entrance. Those in the balconies were not so lucky. The stairs leading down were so narrow and twisting that it was difficult for two people to pass at the best of times. Now, as the crowd panicked and pushed, the entrances were soon blocked. People shed their coats and umbrellas as they ran, creating further chaos.

Then the lights went out. A stagehand, fearing an explosion, had cut off the supply of gas from backstage. In any other theater, his action might have saved lives. The officials at the Ring Theatre, however, had ignored the law instructing them to light oil-fueled lamps at the exits. There were no such lamps, and no lights to guide the audience members outside. Some, stumbling through the dark into dead-end passageways or convoluted exit routes, were killed by smoke and fumes in less than five minutes. Others remained trapped in the balconies, their escape routes blocked by fallen bodies. A few mothers, seeing no way to escape the upper levels, threw their children to the ground floor where an area free of flames still remained. No one survived the fall.

When firefighters finally arrived, they were badly equipped— they hadn't realized the severity of the fire. They had one steam-powered fire engine and a number of hand pumps, and none of their hoses could reach the upper stories. There was basically no hope of extinguishing the blaze. It had already been raging for 25 minutes by that time, and the theater looked like a roaring furnace. Crowds of escapees, including a few half-dressed actors, were gathered on the street to watch in horror.

A group of audience members had found their way to the outside balcony on the second floor, above the theater entrance. There, they were trapped. The fire department had no ladders long enough the reach them, but circles of firemen quickly spread blankets below. One at a time, several hundred audience members leapt, falling more than 20 meters (60 feet) before they were caught and ushered to safety.

One group of firemen was approached by a frantic young woman. Screaming and crying, she pulled at their arms, urging them into the blaze itself. The firefighters thought she must be hysterical. Surely all the audience members had already escaped or were being rescued from the balcony above. There couldn't be anyone still inside the corridors. They restrained the hysterical woman and escorted her to a waiting ambulance.

That hysterical woman was Fraulein Prawlik, the daughter of a Vienna politician, and she was right. There *were* audience members trapped inside—hundreds of them. By the time other survivors convinced the firefighters, it was almost too late to enter the building—thick smoke and unbearable heat blocked the way. The firefighters attempted one foray inside and found a single man still alive, choking on the smoke. After that, it was no longer safe for rescuers to enter.

At 10 p.m., three hours after the fire began, a gas tank exploded and the theater roof collapsed in a booming shower of flame and debris. Transfixed, many local residents remained around the theater for the rest of the night as the roaring flames slowly calmed, blazing walls became smoldering ruins, and the billowing chimney of smoke finally dissipated. By dawn, the fire was over.

The *Wiener Allegemeine Zeitung* newspaper had offices just across the street from the Ring Theatre. The morning news on December 9 read:

> We write with death in our hearts, unable to find words to describe the pain we suffer. A terrible calamity befell our city to-day. The Ring Theatre, formerly known as the Opera Comique, is in ashes, and hundreds of human beings are victims of the catastrophe.

For days, no one knew exactly how many people had died. The building was so unsafe that firefighters had to wait until supports were built against the walls. Only then could they enter and begin to collect the bodies.

As the recovery effort got under way, a reporter stopped one of the firefighters. "How many persons do you think are in the fourth gallery?" he asked.

The firefighter answered: "Two hundred, and possibly five hundred. Who can tell? They are there in thick masses. We shall not know how many people are dead until the people of Vienna have sent in their list of the missing ones to the police."

Every day, the newspapers reported a new death toll. At first, three hundred were thought to be dead. Then four hundred. Then five hundred. Many of the bodies were unrecognizable. Some of the aristocrats from the upper balconies were identified only by the diamond rings and jewels they wore.

When at last the theater was cleared, officials concluded that 794 people had died. The dead included three professors, a member of Parliament, three chorus members, two bank officials, 20 members of the Fruit Exchange, a group of medical students, and hundreds of others. It was as if the fire had torn through the city of Vienna itself, choosing victims from every neighborhood and leaving the entire city in shock.

No Encores, Please

News of the blaze traveled quickly around the world, and offers of help poured in. Many, including the emperor of Austria, donated money to the families of the victims, while other people worked to ensure that such a fire would never happen again. All the theaters in Austria and many across Europe were closed while inspectors reviewed fire precautions, determined to prevent another tragedy. Even 50 years later, when Britain published new guidelines for theater safety, officials mentioned the Ring Theatre fire as one of the main reasons for reform.

In the months after the fire, the theater managers and city officials were taken to court and found guilty of neglect. But their conviction didn't ease the grief of those who had lost friends and family members. Survivors such as Fraulein Prawlik would remember the choking smoke, the screaming victims, and the crushing crowd for the rest of their lives.

Introduction: The Myths and Reality of Fire

Bonaparte, Napoleon. *Napoleon's Letters to Marie Louise.* New York: Farrar & Rinehart, 1935.

Goodman, Edward C. *Fire!* New York: Black Dog & Leventhal Publishers, 2001.

Newton, David E. *Encyclopedia of Fire.* Westport, CT: Oryx Press, 2002.

Fleeing the Flames

"Julia Lemos." The Great Chicago Fire and the Web of Memory website. http://www.chicagohs.org/fire/intro.

Miller, Ross. *The Great Chicago Fire.* Chicago: University of Illinois Press, 1990.

Sawislak, Karen. *Smoldering City.* Chicago: University of Chicago Press, 1995.

Struggle for the Jungle

Mydans, Seth. "Southeast Asia Chokes on Indonesia's Forest Fires." *New York Times,* September 25, 1997, p. A1.

Thoenes, Sander. "In Asia's Big Haze, Man Battles Man-Made Disaster." *The Christian Science Monitor International,* October 28, 1997, p.1.

A Nuclear Nightmare

Cheney, Glenn Alan. *Journey to Chernobyl: Encounters in a Radioactive Zone.* Chicago: Academy Chicago Publishers, 1995.

Illesh, Andrey. *Chernobyl: A Russian Journalist's Eyewitness Account.* New York: Richardson & Steirman & Black, 1987.

Medvedev, Grigori. *The Truth about Chernobyl.* New York: Basic Books, 1991.

Mould, Richard F. *Chernobyl: The Real Story.* Oxford: Pergamon Press, 1988.

Roche, Adi. *Children of Chernobyl.* London: Fount Paperbacks, 1996.

Factory in Flames

McCracken, Elizabeth. "Out of the Fire." *New York Times Magazine,* December 30, 2001, 19–20.

Martin, Douglas. "Rose Freedman, Last Survivor of the Triangle Fire, Dies at 107." *New York Times,* February 17, 2001, p. B8.

"It's Going to Blow!"

Elliot, Floyd, producer. *"Just One Big Mess": The Halifax Explosion, 1917.* Video. National Film Board of Canada, 1991.

Kitz, Janet F. *Shattered City: The Halifax Explosion and the Road to Recovery.* Halifax: Nimbus Publishing, 1989.

Mahar, James, and Rowena Mahar. *Too Many to Mourn: One Family's Tragedy in the Halifax Explosion.* Halifax: Nimbus Publishing, 1998.

Disaster in the Desert

Hawley, T.M. *Against the Fires of Hell: The Environmental Disaster of the Gulf War.* New York: Harcourt Brace Jovanovich, 1992.

Hirshberg, Charles. "Hell on Earth." *Life,* July 1991, 42–45, 48–50.

Horgan, John. "Burning Questions." *Scientific American,* July 1991, 17, 20, 24.

Kelly, J.M. "Fire over Kuwait." *Popular Science,* September 1991, 62–65.

London Is Burning

Hanson, Neil. *The Great Fire of London: In That Apocalyptic Year, 1666.* New Jersey: John Wiley & Sons, 2002.

Latham, Robert, and William Matthews, eds. *The Diary of Samuel Pepys.* Berkeley: University of California Press, 1972.

Porter, Stephen. *The Great Fire of London.* Phoenix Mill: Sutton Publishing, 1996.

Death Train

Brooke, James. "Final Calls Add to Anguish over Korean Subway Fire." *New York Times,* February 20, 2003, p. A12.

Kim, Jin David. "South Korean Subway Blaze 'Close to Hell.'" *Globe and Mail,* February 19, 2003, p. A3.

Kirk, Don. "Use of Dangerous Materials Cited in Korean Subway Fire." *New York Times,* February 22, 2003, p. A4.

Airship Aflame

Botting, Douglas. *Dr. Eckener's Dream Machine: The Great Zeppelin and the Dawn of Air Travel.* New York: Henry Holt & Company, 2001.

"Hindenburg Disaster." Radio Days website. http://www.otr.com/hindenburg.html.

Payne, Lee. *Lighter Than Air.* New York: Orion Books, 1991.

Sullivan, George. *Famous Blimps and Airships.* New York: Dodd, Mead & Company, 1988.

Topping, Dale. *When Giants Roamed the Sky.* Akron: University of Akron Press, 2001.

Operatic Tragedy

The Globe. "The Theatre Horror." December 10, 1881, p. 2.

————. "Vienna's Woe." December 12, 1881, p. 2.

Index

Read these other exciting books in the **Stories from the Edge** series

Tunnels! by Diane Swanson

People have been tunneling since the Stone Age. These 10 gripping stories of human drama beneath the ground are fast-paced and tension-filled.

Scams! by Andreas Schroeder

Scam artists have been tricking people for a very long time. These 10 dramatic stories explore some of the most outrageous and inventive swindlers of all time.

Escapes! by Laura Scandiffio

History is full of daring escapes. The 10 exhilarating stories in this collection take readers around the world and across the ages.

About the Author

TANYA LLOYD KYI has been interested in fire ever since she and her friend Michelle worked together on a Grade 8 science fair project about lightning. They used a static generator, cotton balls, and a small figurine of a golfer to show how lightning worked. Unfortunately, the cotton-ball clouds turned out to be flammable. The innocent golfer was not only struck by lightning, but also somewhat melted by the flames.

Tanya has managed to survive most of her life without setting any more fires, except for one small incident with the Christmas turkey in 2001. She is the author of *Canadian Girls Who Rocked the World*, *Truth*, *The Crystal Connection*, and *My Time as Caz Hazard*. She lives in Vancouver, B.C.